T0258139

Applications of Radioisotopes

Applications of Radioisotopes

Edited by **Peggy Sparks**

New York

Published by NY Research Press,
23 West, 55th Street, Suite 816,
New York, NY 10019, USA
www.nyresearchpress.com

Applications of Radioisotopes
Edited by Peggy Sparks

© 2015 NY Research Press

International Standard Book Number: 978-1-63238-052-4 (Hardback)

This book contains information obtained from authentic and highly regarded sources. Copyright for all individual chapters remain with the respective authors as indicated. A wide variety of references are listed. Permission and sources are indicated; for detailed attributions, please refer to the permissions page. Reasonable efforts have been made to publish reliable data and information, but the authors, editors and publisher cannot assume any responsibility for the validity of all materials or the consequences of their use.

The publisher's policy is to use permanent paper from mills that operate a sustainable forestry policy. Furthermore, the publisher ensures that the text paper and cover boards used have met acceptable environmental accreditation standards.

Trademark Notice: Registered trademark of products or corporate names are used only for explanation and identification without intent to infringe.

Printed in the United States of America.

Contents

Preface

This book is an essential read on radioisotopes. Radioisotopes can be utilized as an alternative for heat conversion into electricity. This is an extensive book which covers various topics related to the physical aspects of radioisotopes and their various applications in some physical and chemical processes. This book will appeal to a broad spectrum of readers interested in the subject of radioisotopes.

Significant researches are present in this book. Intensive efforts have been employed by authors to make this book an outstanding discourse. This book contains the enlightening chapters which have been written on the basis of significant researches done by the experts.

Finally, I would also like to thank all the members involved in this book for being a team and meeting all the deadlines for the submission of their respective works. I would also like to thank my friends and family for being supportive in my efforts.

Editor

Radioisotopes and Some Physical Aspects

Natural Occurring Radionuclide Materials

Raad Obid Hussain and Hayder Hamza Hussain
College of Science/Kufa University
Iraq

1. Introduction

The stellar material, from which the earth was formed, about 4.5 billion years ago, contained many unstable nuclides (Scholten and Timmermans, 1996). Some of the original primordial nuclides, whose half-lives are about as long as the earth's age, are still present. Radiation comes from outer space (cosmic), the ground terrestrial, and even from within over bodies. It present in the air we breathe, the food we eat, the water we drink and in the construction materials used to build our houses. So, radiation is all around us, it is naturally in our environment and it has been since the birth of our planet (Maher and Raed, 2007). Radioactivity of soil environment is one of the major sources of exposure to human (Abusini, 2007).

The ^{235}U, ^{232}Th series and natural ^{40}K are the main source of natural radioactivity in soil (Yasir et al., 2007; Vosniakos et al., 2002). Since these natural occurring radio nuclides materials, (NORMs) such as ^{238}U, ^{232}Th, ^{235}U, and ^{40}K have very long half- lives (up to 10^{10} years), their presence of soils and rocks can simply be considered as permanent. The geological and geographical conditions are the major factors effects, the natural environmental radioactivity and the associated external exposure due to gamma radiation. Thus these radiation levels appear at different levels in the soil of each region in the world (UNSCEAR, 2000).

Issue in terms of radiological protection exposure to natural source of radiation becomes an important. In 1992 the national radiological protection Board (NRPB), estimated that radon accounts for approximately 50% of annual dose of radiation from all sources in the most of the world (Ibrahim, 1999).

An average person receives a radiation dose of about 300 millirem per year from natural sources compared to a dose of about 50 millirem from produced material source of radioactive materials such as medical x-ray (UNSCEAR, 1988). Exposure of public to radiation from any sources is unlikely. The European committee has issued a draft proposal for revision of the basic safety standards for the protection of workers and the general against the dangers of ionizing radiation (Marcelo and Pedro, 2007).

The United Nations Scientific Committee on the Effects of Atomic Radiation established that the world mean dose from natural radiation sources of normal area is estimated to be 2.4 mSv.y^{-1} while for all man- made sources including exposure, is about 0.8 mSv.y^{-1} (UNSCEAR, 1993; Valter at el., 2008). Thus 75% of the radiation dose received by humanity is come from natural radiation source. It is clear that the assessment of gamma radiation dose from natural source is of particular importance as natural radiation is the largest contributor to the external dose of the world population (UNSCEAR, 1988).

Since predominate part of the environmental radiation is found in the upper soil layer, this knowledge ensures radiological control. The ^{238}U, ^{232}Th and ^{40}K have a non-negligible radioactivity (WHO, 1993; José at el., 2005). The high radioactivity of ^{226}Ra and ^{228}Ra and their pressure in soil require particular attention. It is known that even a small amount of a radiation substance may produce a damaging biological effects and that ingested and inhaled radiation can be a serious health risk (Rowland, 1993).

The radiological impact of natural radio nuclides is due to the gamma ray exposure of the body and irradiation of lung tissue from inhalation of radon and its daughters. In general exposure to ionizing radiation often comes from medical diagnosis and therapy application in just food and air and environmental sources. The radiation from the last one cannot be switched of thus the environmental radioactivity surveillance becomes, therefore a necessity. Generally, in Iraq and especially in the middle region area, there is a lack of the scientific information on radioactivity contents of naturally occurring radioactive materials in soil especially in terms of environmental radiological studies.

Based on these facts, one can certify that the knowledge of natural occurring radionuclide materials (MORMs), such as ^{238}U, ^{232}Th and ^{40}K, is an important pre-requisite for evaluation of the rate of exposure absorbed dose by the population in order to estimate their radiological impacts and to establish a data base which will be used as reference to radiation observer in the studied area (NCRP, 1987).

2. Background

According to the source of radiation, radioactivity in the environment may be classified into two general categories; artificial and natural. Natural radioactivity comes from naturally-occurring, Uranium, Thorium and Actinium radioactive series as well as from radioisotopes like Rubidium-87, Indium-115, Lanthanum-138, Neobynium-144, Samarian-147, Luteium-176, Hafinium-174, Vanadium-150, Gadolinium-152, Platinum-190 and 192, Rhenium-187 and Potassium-40. Except for K-40 these non-series radioisotopes occur scarcely. In contrast, K-40 is ubiquitous (Vosnikos et al., 2003). In the other hand artificial activity arises mainly from discarded sources, radioactive wastes, and radioactive fallout in the nature.

There are four distinct natural series: Uranium, Actinium, thorium, and Neptunium as listed in (Table 1). Only uranium, actinium, and thorium series are found in natural. Since the isotope ^{237}Np has a half-life much shorter than the age of the earth (about 5 billion of years), virtually all neptunium decayed within the first 50 millions of years after the earth formed.

Only uranium, actinium, and thorium series are found in natural. Since the isotope ^{237}Np has a half-life much shorter than the age of the earth (about 5 billion of years), virtually all neptunium decayed within the first 50 millions of years after the earth formed.

Series	First Isotope	Half-life(years)	Last Isotope
Uranium	^{238}U	4.5×10^9	^{206}Pb
Actinium	^{235}U	7.10×10^8	^{207}Pb
Thorium	^{232}Th	1.39×10^{10}	^{208}Pb
Neptunium	^{237}Np	2.14×10^6	^{209}Bi

Table 1. Natural series of uranium, actinium, thorium and neptunium

2.1 The NORM decay series

Uranium and thorium are not stable; they decay mainly by alpha-particle emission to nuclides that themselves are radioactive. Natural uranium is composed of three long lived isotopes, ^{238}U, a smaller proportion of ^{235}U and an even smaller proportion of ^{234}U, the decay-series daughter of ^{238}U. Natural thorium has one single isotope, ^{232}Th. Each of these nuclides decays to an unstable daughter leading, in turn, to a whole series of nuclides that terminate in one or other of the stable isotopes of lead. Under normal circumstances, in a natural material, the $^{235}U/^{238}U$ ratio will be fixed and all nuclides in each of the series will be in equilibrium.

Gamma spectrometry of materials containing these nuclides can only be effectively done with a detailed understanding of the decay chains of the nuclides involved.

2.1.1 Uranium series

The products of the decay are called radioactivity series. This series starts with the Uranium-238 isotope, which has a half-life 4.5×10^{10} year as shown in Figure 1 (Henery and John, 1972; Littlefield and Thorley, 1974). Since nuclides have very long half-life, this chain is still present today. The radionuclide ^{238}U decays into ^{234}Th emitting an alpha-particle, the newly formed nuclide is also unstable and decay further (Figure 1). Finally, after total of 14 such steps, emitting 8 alpha particles and 6 Beta particles, accompanied by gamma radiation, stable lead is formed. This series is said to be in secular equilibrium because all their daughters following ^{238}U have shorter half-life than the parent nuclide ^{238}U (Benenson, 2002).

This decay series includes the ^{226}Ra which has half-lives of 1600 year and chemical properties clearly different from those of uranium. ^{226}Ra decay into ^{222}Rn which is an inert noble gas that not form any chemical bonds and can escape into the atmosphere and attacks rapidly to aerosols and dust particles in the air deposited. The radiation emitted at the decay of these products, can cause damage to the deep lungs.

2.1.2 Actinium series

It is also known as Uranium-235 series and starts with ^{235}U and by successive transformations and up in a stable lead ^{207}Pb. It comprises 0.72% of natural uranium. Although only a small proportion of the element, its shorter half-life means that, in terms of radiations emitted, its spectrometric significance is comparable to ^{238}U. The decay series, shown in (Figure 2), involves 12 nuclides in 11 decay stages and the emission of 7 alpha particles (ignoring a number of minor decay branches). Since its abundance is very small, it dose not taken into account in the measurements (Harb, 2004).

Within this series, only ^{235}U itself can readily be measured, although ^{227}Th, ^{223}Ra and ^{219}Rn can be measured with more difficulty. Even though the uncertainties may be high, measurement of the daughter nuclides can provide useful support information confirming the direct ^{235}U measurement or giving insight into the disruption of the decay series.

2.1.3 Thorium series

Natural thorium is 100% ^{232}Th. The decay series is shown in (Figure 3). Six alpha particles are emitted during ten decay stages. Four nuclides can be measured easily by gamma spectrometry: ^{228}Ac, ^{212}Pb, ^{212}Bi and ^{208}Tl. The decay of ^{212}Bi is branched – only 35.94% of decays produce ^{208}Tl by alpha decay. The beta decay branch produces ^{212}Po that cannot be measured by gamma spectrometry. If a ^{208}Tl measurement is to be used to estimate the thorium activity, it must be divided by 0.3594 to correct for the branching (Harb, 2004).

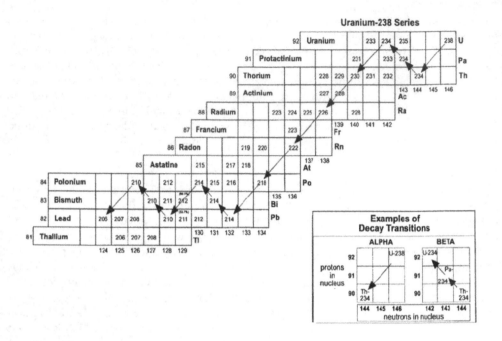

Fig. 1. Schematic diagram of the Uranium-238 series.

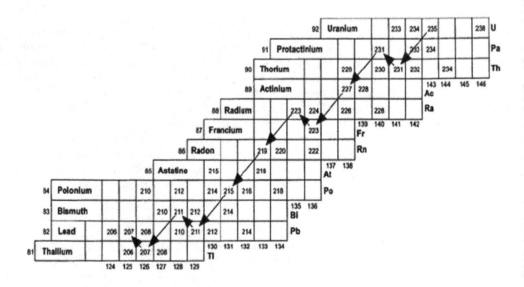

Fig. 2. A schematic diagram of Uranium-235 series (actinium).

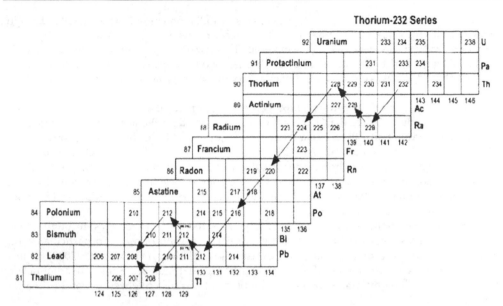

Fig. 3. A schematic diagram of the Thorium-232 series.

2.1.4 Potassium radionuclide

In 1905, J.J. Thompson discovered the radioactivity in ^{40}K is what makes everybody radioactive, it is present in body tissue. This radionuclide can be decayed by three general modes:

a.   Positron emission.

b.   K- electron capture.

c.   Beta emission.

In first mode, ^{40}K radionuclide disintegrates directly into the ground state of ^{40}Ca by the emission of Beta- particle of energy 1321 keV in probability of 88.8% of the decays and no gamma emission is associated with this type of formation (Podgorsak, 2005).

Through the second mode, ^{40}K nuclide can be transformed into stable state (ground state) of ^{40}Ar by two ways, in the first one, ^{40}K disintegrates directly with one jump into ground state of ^{40}Ar with sixteen hundredths of the decays go by electron capture. In the second way, ^{40}K nuclide can be decayed indirectly into the ground state of ^{40}Ar by two stages. firstly, ^{40}K decay into the first excited state of ^{40}Ar. Secondly, the excited nuclide ^{40}Ar, decayed into ground state, accompanied by gamma radiation of 1460 keV energy in probability of 11% of the ^{40}K atoms undergo this change. In the last one (beta emission), a proton will be decayed into positron and ^{40}K changed into ^{40}Ar by probability of 0.0011%.

2.2 Biological effects of radiation

The study of the biological effects of radiation is a very complex and difficult task for two main reasons.

1.   The human body is a very complicated entity with many organs of different sizes, functions, and sensitivities.

2.   Pertinent experiments are practically impossible with humans.

The existing human data on the biological effects of radiation come from accidents, through extrapolation from animal studies, and from experiments in vitro. How and why does radiation produce damage to biological material? To answer the question, one should consider the constituents and the metabolism of the human body. In terms of compounds, about 61 percent of the human body is water. Other compounds are proteins, nucleic acids, fats, and enzymes. In terms of chemical elemental composition, the human body is, by weight, about 10 percent H, 18 percent C, 3 percent N, 65 percent 0, 1.5 percent Ca, 1 percent P, and other elements that contribute less than 1 percent each. To understand the basics of the metabolism, one needs to consider how the basic unit of every organism, which is the cell, functions.

The understandings of natural radiation concepts are essential for radiation protection purpose. The presences of radionuclides in soil affect the common people immensely. Since, the natural radionuclides form 10% of the average annual dose to the human body from all other types of radiation (UNSCEAR, 1993) and exposure to ionizing radiation, in generally considered undesirable at all levels.

Researchers drew attention to the low level exposure; there are three ways, through which the radio nuclides enter the human body: (1) direct inhalation of air born particulates, (2) ingestion through the mouth and (3) entry through the skin (Dipak et al., 2008). Direct exposure to skin is also responsible for radioactive contamination. Some of radionuclide which inters the lung by inhalation affects the blood. Their effectiveness, depend primarily upon two factors:
1.   Kind of the radionuclide.
2.   Physiological of the exposed person

The effects of radioactive in take depend upon the physical and chemical form and the root through which the radionuclide inter the body. These effects may cause damage to genetic organs, and eye defects and skin smear and destroy the circulatory system and lung cancer.

Exposure to low radiation ray lead to somatic infirmities like cancer and genetic defects such as mutation and chromosome aberrations. Gene modifications may result such conditions and diseases as asthma, diabetes, anemia. Genetic changes are passed on from one generation to another (Gerrado, 1974).

When people are exposed to certain levels of ^{238}U, ^{232}Th and ^{40}K for a long period of time cancer of the bone and hazard cavity may result (Nour, 2004). When radium inters the body by ingestions and inhalation, its metabolic behavior in similar to that of calcium faction of it will be deposited in bone where the remaining fraction being distributed uniformly in the soft tissues, thus the most radiotoxic and most important, among the several radionuclide in the radioactive decay chain the two natural series of uranium and thorium are ^{226}Ra and ^{228}Ra. The biological radiation effectiveness can be dividing into two types:

2.2.1 Body effectiveness
Cell is the basic unit of living tissue. Cells are complex structures enclosed by a surface membrane. DNA (deoxyribonucleic acid) is existed in the central of nucleus and considered as code of the structure, function, and replication of the cell. The famous "double helix" of the DNA molecule has a diameter of about (2nm). The induction of cancer or of hereditary disease by low levels of ionizing radiation is believed to be related to damage of the DNA molecules. This can be happen direct by ionization of the molecule, or indirectly through ionization of the water molecules in the cell (Cottingh and Greenwood, 2001). A single broken start and of DNA is rapidly repaired by cellular enzyme system, the unbroken strand of the DNA acting as template.

The water represents nearly 80% of the human body. When a body exposed to an ionized radiation, the large effect will be happened on the water molecules. The break up of water molecule may Produce (. OH^- .) on that is highly reactive chemically and may attack the DNA molecule.

If, at the same time, there is adjacent damage to the other cell that may be errors in the repair process. The cell may die or damage to cell which cause later uncontrolled cell division. The incident radiation on the body, the water molecule will be ionized the water molecule and free electron will be liberated. The mechanism of this process can be summarized in the following equations.

$$H_2O \longrightarrow H_2O^+ + ^-e$$

($^-$e) is called dried electron, its energy proportional with the incident radiation energy. This dried electron will be losses its energy through its path in the body and caused some ionizing and excitation of other atoms and nuclides.

At low energy some water molecules shall be capture these dried electrons produced water electron which denoted by e- symbol. The velocity of water electron e- in the human cells is less than of dried electrons by factor of 105 times (Podgorsak, 2005).

The water positive ion (H_2O^+) will be interacts with the free Hydrogen according to the following equation:

$$H_2O^+ + H_2O \longrightarrow .OH + H_3O^+$$

The water electron e- will be interacting with H_3O^+ reduced water and free hydrogen root ($H_\bullet$):

$$^-e + H_3O^+ \longrightarrow H_2O + H_\bullet$$

It is found that the effectiveness of free hydroxyl root five times of that of free hydrogen root. This free root will be interacting with the organic molecules and other constituent of the cell causing change in their chemical properties which leads to distortion in their functions which may be lead to death. The effect degree of ionizing radiation depends on the type, energy and intensity of radiation and exposure time. The body effective classified into two types:

2.2.2 The early effectiveness

The germ cells in human are the most sensitive in the body if they exposed to radiation, in a dolts these cells presents in double number as compared with children in infancy this increase related to cell division and generation of completely similar new cells.

During division there are spindle-like particulars cells as chromosomes that reveal a hung number of granulated particles with a special arrangement called as genes the later are the responsible for the individually inherited character. the destruction or change of the geneses or chromosomes many reveal a character that not previously present in parents this change called as genetic mutation .

Radiation will increase the probability of occurring of genetic mutation and childhood abnormalities or infection with certain genetic discuses. If the divided cell exposed to radiation, these cells may undergo abnormal cell division, in the same time these cells may have the capacity proof reading of genetic mistakes that faced with hence may repair any

Biological disturbance. The radiation in this harm, but in accordance, they have new characters that may transform to embryos ending with exposed obvious genetic mutations. These genetic mutations not always harmful, in contrast, may result in favorable characters like the gain a good quality fruits in shape or size in the same time, animal's cell if exposed to genetic mutation may result in improved characters.

This type take place when the whole body expressed to radiation for high dose through short time some results can be appear in a few days or weeks, likes, reducing in weigh change in the blood cells hair loss and redness of the skin, and sometime the death is probable.

3. Experimental study

3.1 Gamma spectrometer

In this study we used gamma spectroscopy to determine the ^{238}U, ^{232}Th and ^{40}K in surface soil layer around the uranium mine at Najaf city.

Gamma spectroscopy is one of many famous techniques are used to measure the NORMs contents in the different environmental elements. It has many advantages such as high accuracy, measure wide energy range and different type samples and not need a chemical method in sample preparation. Beside these advantages gamma spectrometry of NORM is still difficult for a number of reasons. First, the activity levels are low and, if statistically significant results are to be obtained, need long count periods, ideally on a gamma spectrometer whose construction and location are optimized for low activity measurements. The second difficulty is the matter of spectrometer background (i.e. a large number of peaks that one might see in background spectra). Many of these are due to the NORM nuclides in the surroundings of the detector. Any activity in the sample itself must be detected on top of all that background activity. In many cases, it will be necessary to make a peaked-background correction in addition to the normal peak background continuum subtraction. All of those difficulties are then compounded by the fact that there are a large number of mutual spectral interferences between the many nuclides in the decay series of uranium and thorium.

The gamma rays levels were measured by integral counting using a spectrometer consist of a scintillation detector NaI(Tl) of ($2'' \times 2''$) crystal dimension with resolution value of 6.48% for line energy of 662 keV, scalar, shielding and specially designed sample container that allowed the sample to surround the scintillation detector at the top and on the sides. This system was computer controlled. The detector was connected to the amplifier through preamplifier unit; an analog to digital converter (ADC) of 4096 channels was assembled to the system. The spectroscopic measurements and analysis were performed via the CASSAY software into the PC of the laboratory.

In order to reduce the background radiation due to different radiation hazard, the detector was maintained in vertical position and shield by a cubic chamber of two layers starting with copper of 2mm thick followed by lead of 10 cm thick. The cosmic rays, photons and electrons, are reduced to a very low level by the 10 cm of lead shielding. This interaction will produced x-ray with low energy which can be suppressed by the copper layer (Aziz, 1981). The x-rays can be also come from radioactive impurities like antimony in the lead.

The spectrometer was calibrated for energy by acquiring a spectrum from radioactive standard sources of known energies like ^{60}Co (1332 keV, 1773 keV) and ^{137}Cs (662 keV). To measure the counting efficiency of the system, ^{22}Na, ^{57}Co, ^{60}Co, ^{109}Cd, ^{133}Ba and ^{137}Cs standard sources of gamma rays were used (Table 2). The relative intensities of the photo-peaks corresponding to their gamma rays lines have been measured.

Isotope	E (keV)	I %	Isotope	E (keV)	I %
^{22}Na	1274.5	99.95	^{133}B	80.99	34
^{57}Co	122.1	85.6		276.39	7.16
	136.4	10.88		302.85	18.3
^{60}Co	1173.2	99.97		356	62
	1332.5	99.98		383.85	8.9
^{109}Cd	88.03	3.6	^{137}Cs	661.6	85.1

Table 2. Energies and transition probabilities of standard sources (Heath, 1997)

3.2 Study area and sampling

Twenty five soil samples were collected in area of approximately 40000 m², located around the uranium mine at Najaf governorate. The latitude and longitude of this area are 31° 52′ 254″ N and 22° 26′ 221″ E. We used systematic grid sampling system involves subdividing the area of concern by using a square and collecting samples from the nodes (intersections of the grid lines). The origin and direction for placement of the grid is done, where the mine was centred in grid. From that point, a coordinate axis and grid is constructed over the whole site. The distance between sampling locations was 50 m (Figure 4). Systematic grid sampling is often used to delineate the extent of contamination and to define contaminant concentration gradients (IAEA, 2004).

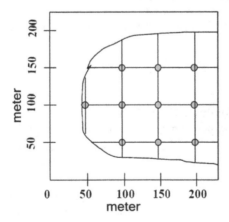

Fig. 4. Systematic grid sampling method

In order to measure the NORMs in soil surface, 25 soils samples were collected, one sample average from each point, was taken by digging a hole at a depth of 35 cm before the ground surface. The soil texture for all samples was very similar. The collected samples were transferred to labeled closed polyethylene bags and taken to the laboratory of radiation detection and measurement in the physics department, collage of science, university of Kufa. In this work a 1.4 litter polyethylene marinelli beaker was used as a sampling and measuring container. Before use, the containers were washed with dilute hydrochloric acid and rinsed with distilled water. The soil samples were prepared for analysis by drying, sieving and kept moisture free by keeping 24 hours in the oven at 100C°. They were mechanically crushed and sieved through of 0.8 mm pore size diameter sieved to get homogeneity (R2).

To remove completely the air from sample, the later was pressed on by light cap of the marinelli beaker. The respective net weights were measured and record with a high sensitive digital weighting balance with a percent of ±0.01%. After that about 1 kg of each sample was then packed in a standard marinelli beaker that was hermetically sealed and dry weighted. The sample was placed in face to face geometry over the detector for along time measurement.

3.3 Specific activity

The definition of activity refers to the number of transformations per unite time. Since the fundamental unite time is the second the quantity activity is measured in disintegrations per second or (dps). In 1950, the international joint commission on standards Unit and constants of radioactivity define the curie by accepting 37 billion dps as curie of radioactivity regardless of its source or characteristics. The SI derived unit of activity is the Becquerel (Bq) and is that quantity of radioactive material in which one atom is formed per second or under goes one disintegration per second (1 dps).

The activity of ^{238}U was estimated from the 1765 keV gamma transition energy of ^{214}Bi (17% possibility). Also the activity of ^{232}Th was measured from the 2614 keV gamma transition energy of ^{208}Tl (100% possibility) whereas ^{40}K activity was determined using the 1460 keV gamma ray line (10.7% possibility) (Vosniakos, 2003).

The specific activity is defined as activity per unite mass of radioactive substance and the reported in units such as Curie per gram or Becquerel per kilogram (Bq/kg). The specific activity of each radionuclide was calculated using the following equation (UNSCEAR, 2000).

$$A = \frac{A'}{\varepsilon \cdot I_\gamma \cdot m \cdot t}$$

Where A the specific activity of the radionuclide in Bq/kg, A′ the liquid count, ε the counting efficiency, I_γ the percentage of gamma emission probability of the radionuclide under study, t the counting time in second and m the mass of the sample in kg.

3.4 Radium equivalent activity

To represent the activity levels of ^{238}U, ^{232}Th and ^{40}K which take into account the radiological hazards associated with them, a common radiological index has been introduced. This index is called radium equivalent activity (Ra_{eq}) and is mathematically defined by (UNSCEAR, 2000).

$$Ra_{eq}(Bq/kg) = A_U + 1.43A_{Th} + 0.077A_K$$

Where A_U, A_{Th} and A_K are the specific activities of Uranium, Thorium, and potassium respectively. This equation is based on the estimation that 10 Bq/kg of ^{238}U equal 7 Bq/kg of ^{232}Th and 130 Bq/kg of ^{40}K produced equal gamma dose. The maximum value of Ra_{eq} must be less than 370 Bq/kg. Also the Ra_{eq} value of 370 Bq/kg is equivalent to the annual dose equivalent of 1.5 mSv/y, which we assumed to be the maximum permissible dose to human from their exposure to natural radiation from soil in one year.

3.5 Absorbed does rate in air

Absorbed dose rate defined as the ratio of an incremental dose (dD) in a time interval (dt).

$$AD = \frac{dD}{dt}$$

Gamma dose rate in air, one meter above the ground, is used for the description of terrestrial radiation, and is usually expressed in nGy/h or pGy/h. the absorbed dose rate due to gamma radiation of naturally occurring radionuclide (^{238}U, ^{232}Th, and ^{40}K), were calculated on guidelines provided by (UNSCEAR, 2000).

$$AD(nG/h) = 0.462A_U + 0.621A_{Th} + 0.0417A_K$$

Where 0.462, 0.621 and 0.0417 are the conversion factors for ^{238}U, ^{232}Th and ^{40}K assuming that the contribution natural occurring radionuclide can be neglected as they contribute very little to total dose from environmental background.

3.6 Annual effective doses

To estimate annual effective doses, account must be taken of (a) the conversion coefficient from absorbed dose in air to effective dose and (b) the indoor occupancy factor. The average numerical values of those parameters vary with the age of the population and the climate at the location considered. In the UNSCEAR 1993 Report, the Committee used 0.7 Sv.Gy/y for the conversion coefficient from absorbed dose in air to effective dose received by adults and 0.8 for the indoor occupancy factor, i.e. the fraction of time spent indoors and outdoors is 0.8 and 0.2, respectively. These values are retained in the present analysis. From the data summarized in this Chapter, the components of the annual effective dose are determined as follows: (UNSCEAR, 1993).

Indoor (nSv) = absorbed dose nGy/h × 8760 h × 0.8 × 0.7 SvG/y

Outdoor (nSv) = absorbed dose nGy/h × 8760 h × 0.2 × 0.7 SvG/y

The resulting worldwide average of the annual effective dose is 0.48 mSv, with the results for individual countries being generally within the (0.3 - 0.6) mSv range. For children and infants, the values are about 10% and 30% higher, in direct proportion to an increase in the value of the conversion coefficient from absorbed dose in air to effective dose.

3.7 Hazard index

To reflect the external exposure, a widely used hazard index, called the external hazard index (H_{ex}), which is defined as following:

$$H_{ex} = \frac{A_U}{370} + \frac{A_{Th}}{259} + \frac{A_K}{4810}$$

There is another hazard index called internal hazard index (H_{in}), which is given by equation.

$$H_{in} = \frac{A_U}{185} + \frac{A_{Th}}{259} + \frac{A_K}{4810}$$

The values of the index must be less than the unity in order to keep the radiation hazard to be insignificant unity corresponds to the upper limit of radiation equivalent activity (370 Bq/kg).

4. Results and discussion

The spectra of twenty five surface soil samples surrounded the abandoned Uranium mine hole have been analyzed. The specific activity of ^{238}U, ^{232}Th, ^{40}K and Radium equivalent activity (Ra_{eq}) are given in Table 3. The specific activity (Bq/kg) varied from 37.31 to 1112.47 (mean = 268.16), 0.28 to 18.57 (mean = 6.68) and 132.25 to 678.33 (mean = 277.49) for ^{238}U, ^{232}Th and ^{40}K respectively.

The obtained results are comparable to the worldwide average recommended by UNSCEAR which are 30, 35 and 400 Bq/kg for ^{238}U, ^{232}Th and ^{40}K respectively (UNSCEAR, 2000). It was found that all values of ^{238}U specific activities are higher than the worldwide average whereas those of ^{232}Th are less than it. For ^{40}K, it is clear that the specific activities, with the exception of five samples, are found to be less than worldwide average.

Obviously demonstrate that the minimum and maximum specific activity values of ^{238}U are least by factor of 4 and 37 higher than the corresponding values obtained worldwide average. The large variation between the specific activities obtained for ^{238}U and other two radionuclides can be easily ascribed to the high content of uranium in the neglected waste of drilling and exploration operations on the surface soil surrounding the mine. The contour maps (radiological maps) of the activity distribution of ^{238}U, ^{232}Th and ^{40}K in the study area are shown in Figures 5, 6 and 7. From Figure 5, we can observe three regions with a highest specific activity values of ^{238}U situated at northeast, east and south-west portions of the hole mine. In contrast, Figure 7 indicates that high concentrations of ^{40}K occupies the same positions of ^{238}U while for ^{232}Th there are no placements have activities require attention as shown in Figure 6.

The calculated Ra_{eq} values for all samples were also presented in Table 3. It may be seen that Ra_{eq} oscillates between 52.727 and 1189.845 with an average of 299.09 Bq/kg. It is observed that the values of Raeq in twenty one samples were less than the acceptable safe limit of 370 Bq/kg (OECD, 1979; UNSCEAR, 1982; UNSCEAR, 1988). As shown in Table 3 there are four values greater than worldwide average. As a rule, the matter whose Ra_{eq} exceeds 370 Bq/kg is discouraged (Beretka and Mathew, 1985). Figure 8 demonstrates the distribution of Ra_{eq} and it appears three positions have highest values.

The calculated absorbed dose rate of samples was listed in Table 3. As shown in Table 3, the values ranged from 25.02 to 553.01 with an average value of 139.61 nG/h which is nine fold higher than the world average of 15 nG/h recommended by UNSCEAR (UNSCEAR, 2000). It can be seen that all values were much higher than the world average.

Sample code	specific activity (Bq/kg)			Ra_{eq} (Bq/kg)	AD nG/h
	$238U$	$232Th$	$40K$		
S11	72.17±4.43	3.38±0.69	253.19±8.79	96.49	46.00
S12	59.37±4.02	0.28±0.20	184.47±7.51	73.97	35.29
S13	213.24±7.62	5.63±0.89	238.84±8.54	239.68	111.97
S14	39.76±3.29	0.28±0.20	174.70±7.30	53.61	25.82
S15	480.12±11.43	8.02±1.06	415.06±11.26	523.54	244.10
S21	138.07±6.13	2.81±0.63	176.23±7.34	155.65	72.88
S22	645.42±13.26	10.13±1.19	521.04±12.61	700.02	326.20
S23	312.64±9.23	6.47±0.95	409.87±11.19	353.45	165.54
S24	249.18±8.24	4.08±0.76	283.43±9.30	276.83	129.47
S25	37.31±3.19	3.66±0.72	132.25±6.36	52.72	25.02
S31	122.82±5.78	2.39±0.58	153.63±6.85	138.06	64.63
S32	285.95±8.82	12.94±1.35	219.60±8.19	321.36	149.30
S33	122.82±5.78	5.35±0.87	153.52±6.85	142.29	66.46
S34	273.97±8.64	6.47±0.95	211.96±8.05	299.54	139.43
S35	78.43±4.62	4.08±0.76	232.12±8.42	102.13	48.44
S41	142.43±6.23	4.92±0.83	205.85±7.93	165.31	77.44
S42	324.07±9.39	7.03±0.99	264.49±8.99	354.48	165.11
S43	860.29±15.31	13.65±1.39	678.33±14.39	932.04	434.21
S44	175.38±6.91	3.80±0.73	229.67±8.38	198.49	92.96
S45	1112.47±17.41	18.57±1.62	459.96±11.85	1189.84	553.01
S51	167.48±6.75	7.32±1.01	302.06±9.60	201.20	94.51
S52	276.69±8.68	10.41±1.21	156.98±6.92	303.66	140.84
S53	139.16±6.16	7.32±1.01	158.21±6.95	161.81	75.43
S54	85.78±4.83	7.88±1.05	201.27±7.84	112.54	52.91
S55	288.94±8.87	10.27±1.20	320.38±9.89	328.29	153.22
Min.	37.31	0.28	132.25	52.72	25.26
Max.	1112.47	18.57	678.33	1189.84	553.01
mean	268.16	6.68	277.49	299.09	139.61

Table 3. Specific activity, Radium equivalent activity and absorbed dose rate of soil samples.

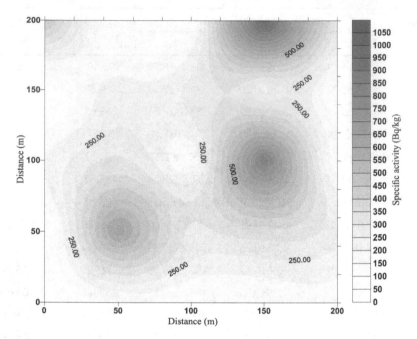

Fig. 5. Specific activity distribution of [238] U.

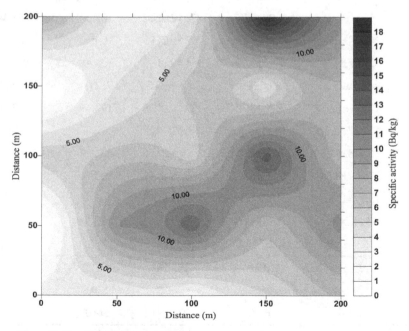

Fig. 6. Specific activity distribution of [232]Th.

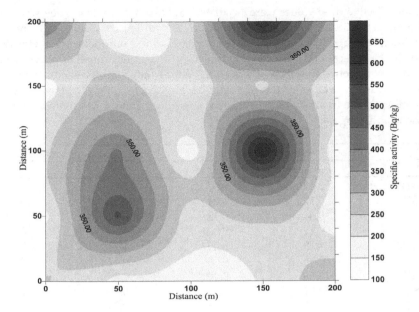

Fig. 7. Specific activity distribution of ^{40}K.

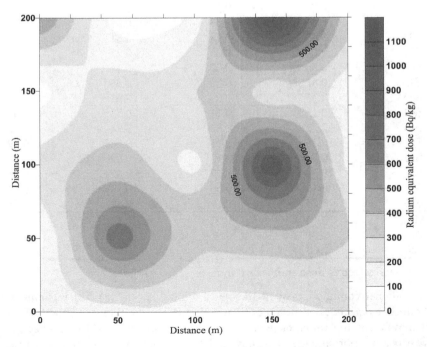

Fig. 8. Distribution of Radium equivalent in surface soil around mine.

The annual effective dose values were calculated and listed in Table 4. They were found to be in the range 0.123 to 2.713 mSv/y with an average value 0.68 mSv/y and from 0.031 to 0.6780 with an average value of 0.17 mSv/y for indoor and outdoor annual effective dose respectively. In general and as shown in Table 4, for indoor annual effective dose, It is important here to notice that there are fourteen sample have values higher than the word average whereas, the values of the rest samples are close or slightly above of the world average value of soil. In other words, all values of outdoor annual effective dose were below the worldwide average.

Sample code	Annual dose (mSv)		Hazard index	
	indoor	outdoor	H_{ex}	H_{in}
S11	0.22	0.05	0.26	0.45
S12	0.17	0.04	0.20	0.36
S13	0.54	0.13	0.64	1.22
S14	0.12	0.03	0.14	0.25
S15	1.19	0.29	1.41	2.71
S21	0.35	0.08	0.42	0.79
S22	1.60	0.40	1.89	3.63
S23	0.81	0.20	0.95	1.80
S24	0.63	0.15	0.74	1.42
S25	0.12	0.03	0.14	0.24
S31	0.31	0.07	0.37	0.70
S32	0.73	0.18	0.86	1.64
S33	0.32	0.08	0.38	0.71
S34	0.68	0.17	0.81	1.55
S35	0.23	0.05	0.27	0.48
S41	0.38	0.09	0.44	0.83
S42	0.81	0.20	0.95	1.83
S43	2.13	0.53	2.51	4.84
S44	0.45	0.11	0.53	1.01
S45	2.71	0.67	3.21	6.22
S51	0.46	0.11	0.54	0.99
S52	0.69	0.17	0.82	1.56
S53	0.37	0.09	0.43	0.81
S54	0.26	0.06	0.30	0.53
S55	0.75	0.18	0.88	1.66
Min.	0.12	0.03	0.14	0.24
Max.	2.71	0.67	3.21	6.22
Mean	0.68	0.17	0.80	1.53

Table 4. Annual effective dose and hazard indexes of soil samples.

The international commission on Radiological Protection (ICRP) has recommended the annual effective dose equivalent limit of 1 mSv/y for the individual members of the public and 20 mSv/y for the radiation workers (ICRP, 1993). The worldwide average annual effective dose is approximately 0.5 mSv and the results for individual countries being generally within the 0.3 to 0.6 mSv range (UNSCEAR, 2000).

In addition, the calculated values of hazard index for the soil samples were ranged from 0.142 to 3.216 with an average value of 0.808 and from 0.243 to 6.222 with an average value of 1.533 for external (H_{ex}) and internal (H_{in}) respectively as mentioned in Table 2.
Out of 25 positions, 4 for H_{ex} and 13 for H_{in}, have values very higher than unity. Since these values are dispersed randomly within a limited area around the min hole, therefore, according to the report of European Commission in Radiation Protection, the area study is not safe and posing significant radiological threat to the population (European Commission, 1999).

5. Conclusion

The surface soil layer around the uranium mine hole has uranium activities greater than worldwide average; this can mainly due to the waste of drilling and exploration left on the surface layer of soil surrounding the mine.
The thorium activities were within normal level in the studied area. Generally, potassium radionuclide in soil samples was in the range of worldwide average.
The absorbed dose rates of studied area are higher than the criterion limit of gamma radiation dose rate with an average of nine times.
Finally, from the radiation protection point of view the studied area is considered to be not safe inhabitants because the values of both internal and external hazard indexes associated with the samples are higher than unity. Thus, the human inside the area are supposed to acquire radiological complication.

6. References

Abusini M. (2007). Determination of Uranium, Thorium and Potassium Activity Concentrations in Soil Cores in Araba valley, Jordan, *Radiation Protection Dosimetry*, 128: 213-216.

Benenson W. (2002). *Hand book of physics*. Fourth Edition. Spriger –Velarg, New York, INC.

Beretka J. and Mathew P. J. (1985). Natural Radioactivity of Australian Building Materials, Industrial Wastes and by Products, *Health Physics*, 48: 87-95.

Cottingh W. N. and Greenwood D. A. (2001). *An introduction to Nuclear physics*, second edition, University of Cambridge, United kingdom.

Dipak G., Argha D., Sukumar B., Rosalima S., Kanchan K. P. (2008). Measurement of Natural Radioactivity in Chemical Fertilizer and Agriculture Soil: Evidence of High Alpha Activity, *Environmental Geochemistry and Health*, 30: 79-86.

European Commission (1999). *Radiological Protection Principle concerning the Natural Radioactivity of Building Materials*. Radiation Protection 112, European Commission ,Brussels.

Gerrado C. Maxino (1974). Radioactive Potassium, *The nucleous*, 12: 4-8.

Harb S. (2004). *On the Human Radiation Exposure as Derived From the Analysis of Natural and Man-Made Radionuclides in Soil*, Ph.D. Thesis, Hannover University.

Henery S. and John R. A. (1972). *Introduction to Atomic and Nuclear Physics*, Fifth edition, Holt, Rineart and Winston INC.

IAEA (2004). *Soil Sampling for Environmental Contaminants*, International Atomic Energy Agency, TECDOC-1415, Vienna, Austria.

Ibrahim N. (1999). Natural Activities of [238]U, [232]Th and [40]K in Building Materials, *J. Environ. Radioact.*, 43: 255-258.

ICRP (1993). International Commission on Radiological Protection, publication 65,Annals of the ICRP 23(2), Pergamon press, Oxford.

José A. dos Santos, Jorge J. R., Cleomacio M. da Silva, Suêldo V. S. and Romilton dos Santos A. (2005), Analysis of the ^{40}K Levels in Soil using Gamma Spectrometry, *Brazilian Archives of Biology and Technology*, 48: 221-228.

Littlefield T.A. and Thorley N. (1974). *Atomic and Nuclear Physics*, Third edition. Van Nostrand Reinhold company, London, UK.

Maher O. El-Ghossain and Raed M. Abu Saleh (2007). Radiation Measurements in Soil in the Middle of Gaza-Strip Using Different Type of Detectors, *The Islamic University Journal (Series of Natural Studies and Engineering)*, 1 5: 23-37.

Marcelo F. M. and Pedro M. J. (2007). Gamma Spectroscopy in the Determination of Radionuclides Comprised in Radioactive Series, *International Nuclear Atomic Conference-INAC*, Santos, SP, Brazil.

NCRP (1987), National Council on Radiation Protection and Measurements, *Exposure of the Population in the United States and Canada from natural background radiation*. No.94, USA.

Nour K. A. (2004), Natural Radioactivity of Ground and Drinking Water in Some Areas of Upper Egypt, *Turkish J. Eng. Env. Sci.*, 28: 345-354.

OECD (1979). *Exposure to Radiation From the Natural Radioactivity in Building Materials*, Report by a Group of Experts of the OECD Nuclear Energy Agency, Organization for Economic Cooperation and Development, Paris, France.

Podgorsak E .B. (2005). *Radiation Physics for Medical Physicist*, Springer Berlin Heidelberg, New York, USA.

Rowland R. E. (1993). Low-Level Radium Retention by the Human Body; A Modification of the ICRP Publication 20 Retention Equation, *Health Phys*, 65: 507-513.

Scholten L. C. and Timmermans. C. W. M. (1996). Natural Radioactivity in Phosphate Fertilizers, *Fertilizer Resrch.*, 43: 103-107.

UNSCEAR (1982). *Sources effects and risks of ionizing radiation*, Report to the General Assembly, With Annexes, United Nations Scientific Committee on the Effects of Atomic Radiation, New York, United Nations.

UNSCEAR (1988). *Sources effects and risks of ionizing radiation*, Report to the General Assembly, With Annexes, United Nations Scientific Committee on the Effects of Atomic Radiation, New York, United Nations.

UNSCEAR (1993). *Sources effects and risks of ionizing radiation*, Report to the General Assembly, With Annexes, United Nations Scientific Committee on the Effects of Atomic Radiation, New York, United Nations.

UNSCEAR (2000). *Sources effects and risks of ionizing radiation*, Report to the General Assembly, With Annexes, United Nations Scientific Committee on the Effects of Atomic Radiation, New York, United Nations.

Valter A. B., Francisco J. F. F., William C. P. (2008). Concentration of Radioactive Elements (U, Th and K) Derived From Phosphatic Fertilizers in Cultivated Soils, *Brazilian Archives of Biology and Technology*, 51: 1255-1266.

Vosniakos F., Zavalaris K., Papaligas T. (2003). Indoor Concentration of Natural Radioactivity and The Impact to Human Health, *Journal of Environmental Protection and Ecology*, 4: 733-737.

Vosniakos F., Zavalaris K., Papaligas T., Aladjadjiyan A. and Ivanova D. (2002). Measurements of Natural Radioactivity Concentration of Building Material in Greece, *Journal of Environmental protection and Ecology*, 3: 24-29.

WHO (1993). *Guideliens for Drinking –Water Quality "Recommendations*, World Health Organization, Geneva.

Yasir M.S., Majid A. Ab., Yahaya R. (2007). Study of Natural Radionuclides and Its Radiation Hazard Index in Malaysia Building Material, *Radioanal. Nucl. Chem.*, 273: 539-541.

2

Diffusion Experiment in Lithium Ionic Conductors with the Radiotracer of [8]Li

Sun-Chan Jeong
Institute of Particle and Nuclear Studies (IPNS)
High Energy Accelerator Research Organization (KEK) 1-1 Oho
Japan

1. Introduction

Radioactive nuclides have been used in materials science for many decades. Besides their classical application as tracers for diffusion studies, nuclear techniques (i.e. Mössbauer Spectroscopy, Perturbed Angular Correlation, β-Nuclear Magnetic Resonance, Emission Channeling, etc.) are now being routinely used to gain microscopic information on the structural and dynamical properties of the bulk of materials via hyperfine interactions or emitted particles themselves (Wichert & Diecher, 2001). These nuclear techniques were primarily developed in nuclear physics for detecting particles or γ-radiations emitted during the decay of the radioactive nuclides. More recently these techniques have also been applied to study complex bio-molecules, surfaces, and interfaces (Prandolini, 2006). With the advent of most versatile 'radioactive isotope beam (RIB) factory' represented by the on-line isotope separator (ISOL)–based RIB facility (see Fig. 1), the possibilities for such investigations have been greatly expanded during the last decade (Cornell, 2003).

At the tandem accelerator facility of Japan Atomic Energy Agency (JAEA)-Tokai, a RIB facility, TRIAC (Watanabe et al., 2007)-Tokai Radioactive Ion Accelerator Complex- is operating since 2005. In the facility, short-lived radioactive nuclei produced by proton or heavy ion induced nuclear reactions can be accelerated up to the energy necessary for experiments. The energy is variable in the range from 0.1 to 1.1 MeV/nucleon, which is especially efficient for studies of the bulk of materials by using the RIBs as tracers. It allows us to implant (incorporate) the RIBs into specimens at a proper depth, avoiding the difficulties caused by the surface (e.g. diffusion barrier like oxide layers that often hampers the incorporation of those radioactive isotope probes into the materials of interest). In the facility, the separation and the implantation of radioactive probes are integrated into one device, as shown in Fig.1. Although the main concerns of the facility are nuclear physics experiments, as an effort to effectively use the available radioactive isotope beams at the TRIAC for materials studies, we have developed a diffusion tracing method by using the short-lived radioactive nuclei of [8]Li as diffusion tracers. The method has been successfully applied to measure diffusion coefficients in a typical defect-mediated lithium ionic conductor (refer to Chandra, 1981 for ionic conductors). We found that the present method is very efficient for the micro-diffusion, where the diffusion length is about 1μm per second.

In the following, the experimental method using ^{8}Li as a diffusion tracer and its application for measuring diffusion coefficients in inter-metallic lithium compounds will be reviewed, and then we will discuss possible extensions of the present method to study lithium diffusion with higher sensitivity such as the diffusion across interface in micrometer scale and the diffusion in nano-scale.

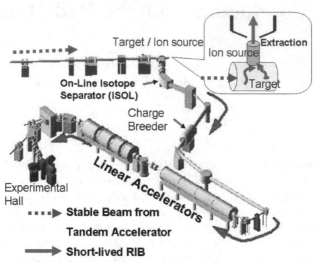

Fig. 1. Layout for ISOL-based RIB production at the TRIAC: Radioactive nuclei (e.g. uranium fission fragments) are produced by nuclear reactions induced in a thick target by a stable beam (e.g. UC$_2$ target by the irradiation of 30-MeV proton from the JAEA tandem accelerator which is not shown here). The target kept at a high temperature (~2000K) permits the fast diffusion of the reaction products into the ion source where they are ionized by plasma impact or surface ionization, as schematically shown in the inset. The singly charged ions (usually positively charged, 1+) are then extracted, mass-separated in a magnetic dipole field of the ISOL, and further, after being boosted to higher charge states by a charge breeder, accelerated to the energies necessary for experiments. For producing the radiotracer of ^{8}Li, the heavy ion beam of ^{7}Li on a ^{13}C-enriched graphite target was used (see the subsection 2-2).

2. Non-destructive on-line diffusion experiment

Over more than half a century, diffusion studies in solids using radioactive tracers (Tujin, 1997) have played an important role in understanding the underlying mechanism of atomic transport in solids, which is of great importance in a number of branches of materials science and engineering. Conventional diffusion studies by means of the radiotracer method in conjunction with a serial sectioning technique have been performed as follows (Wenwer at al., 1996): A small amount of a suitable radioactive isotope of the diffusing element is deposited onto the sample surface of interest. After a diffusion annealing at a temperature of T for a time of t, the sample is sectioned in parallel to the initial surface. An appropriate counting device measures the specific tracer activity in each section, which is proportional to the concentration of the diffusing species. The concentration-depth profile of the tracer is then compared with the solution of Fick's second law under the experimental boundary conditions, yielding the tracer diffusion coefficient at the temperature T in the sample. The choice of an appropriate serial sectioning technique depends on the average diffusion length related to the annealing time and temperature, often by $2(Dt)^{1/2}$, where D is the tracer diffusion coefficient in the sample. The method is consequently destructive.

Although the conventional radiotracer method for diffusion studies has yielded the most accurate diffusion coefficients, the method has not yet been applied for some elements because of no-availability of radiotracers with adequate lifetimes (a rather long lifer time is needed for the process of annealing and sectioning). Among them, ^{8}Li, the radioactive isotope of Li with a half-life of 0.84 s is of special interest for practical issues; how well Li ions move in the secondary Li ion batteries. Fast Li diffusion is desirable in battery materials, i.e., Li ionic conductors for materials of electrodes and solid electrolyte. For studies on the macroscopic diffusivity of Li in Li ionic conductors, various electro-chemical methods (Sato et al., 1997) have been usually adopted up to now. However, the diffusion coefficients are scattered over several orders of magnitude, strongly depending on the method used for the measurement. Therefore, the diffusion coefficients measured in different ways, e.g., by using the radiotracer of Li, are highly required to settle down such disagreements. Such an experimental knowledge on the Li diffusion in as-developed materials for the battery is also of importance in the recent general efforts to design the battery by simulations based on the first principle.

2.1 Principle of the measurement of Li diffusion coefficients with the radiotracer of ^{8}Li

The radiotracer ^{8}Li decays through β-emission to ^{8}Be with a half lifetime of 0.84 s, which immediately breaks up into two α-particles with energies continuously distributed around 1.6 MeV with a full width at half maximum (FWHM) of 0.6 MeV (Bonner et al., 1948).

As for a diffusion tracer, special attention has been paid on the energy loss of the α-particles in the sample of interest, which is sensitive to the diffusion length of about 1 μm. In an ideal case when the radiotracer emits monochromatic α-particles, the amount of incidental energy loss of the α-particles on their passage to the surface of the solid of interest depends on the position of the decaying emitter; the measured energies of the α-particles passed through the solid are closely related to the decaying positions of the tracer. The time evolution of the energy spectra is therefore supposed to be a measure of the diffusivity of the tracer in the solid. The energy spectra are broadening with increasing diffusion time; the tracer diffusion coefficients could be simply obtained by the time-dependent widths of the measured energy spectra if the inherent energy of the emitted charged particles is well defined. In the present case, however, the inherent energy distribution of the α-particles is continuous and broad (Bonner et al., 1948). Although the correspondence between the emitted position and measured energy of the charged particles is not as simple as above, it was shown in the simulation that the tracer diffusion coefficient could be obtained from the time-dependent yields of α-particles emitted by diffusing ^{8}Li with the help of the simulation (Jeong et al., 2003).

Figure 2 shows schematically the principle of the measurement: Implanting the beam of ^{8}Li with a properly adjusted energy into a depth, which is deeper than the average range of α-particles, we can make a situation where most α-particles stop in the sample. After implantation, since the primary implantation profile is broadening by diffusion, the α-particles emitted by ^{8}Li diffused toward the surface can survive and come out of sample with measurable energies. Then a charged particle detector located close to the sample surface could selectively detect α-particles from ^8Li diffusing toward the sample surface, since the implantation-depth is deeper than the average range of α-particles in the present case. Therefore, the temporal evolution of α-particle yields that come out of the sample would be a measure of the diffusivity of Li.

It should be noted that the present diffusion time is different from that of the conventional radiotracer method for diffusion studies (Wenwer at al., 1996) because the tracer in the present method diffuses all the time of the measurement. This is the reason why we call the present method as a non-destructive on-line measurement of diffusion.

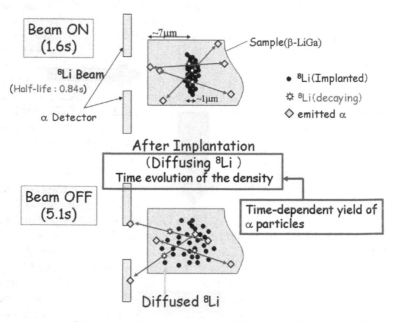

Fig. 2. Schematic view of the principle for measuring diffusion coefficients in β-LiGa (a typical sample of the inter-metallic Li compounds presently used) when the short-lived α-emitting ^8Li was used as the diffusion tracer. The yield of α particles measured at a time is a measure of the diffused distribution of ^8Li primarily implanted in the sample with a depth of about 7 μm.

2.2 Experimental set-up

Along the idea described in the previous subsection, we made an experimental set-up as shown in Fig. 3. All components were installed in a chamber evacuated to 1×10^{-4} Pa.

For producing ^8Li ($T_{1/2} = 838ms$), we have used a neutron transfer reaction of ^{13}C (^7Li, ^8Li), using a sintered target of 99%-enriched ^{13}C. The 99%-enriched ^{13}C graphite disk (~10mm in diameter with a thickness less than 1mm) was mounted to the catcher position of the surface ionization type ion source with a beam window of 3-μm thick tungsten (Ichikawa et al., 2003). The target was bombarded with a 67-MeV $^7Li^{3+}$ beam with an intensity of about 100 pnA (particle nano-Ampere). The produced ^8Li was ionized, mass-separated as a radiotracer beam by the ISOL, and then injected to the post accelerators of the TRIAC, as shown in Fig.1.

Provide by the TRIAC, the radiotracer beam ⁸Li of about 4 MeV with an intensity of about 10^4 particles/s was periodically implanted to a sample of β-LiGa with a following time sequence; 1.6 s for implantation (beam-on) and 5.1 s for subsequent diffusion (beam-off). With the incident energy, the ⁸Li radiotracer can be implanted into the implantation-depth of about 7 μm from the front surface of the sample of β-LiGa. The α particles coming out of the sample were measured as a function of time by an annular solid-state detector (SSD) installed close to the front surface of the sample as shown in Fig. 3. The sequence was repeated to obtain good statistics, where the time-zero was always at the beginning of the implantation. Before starting the measurement, the sample was set at a temperature where the diffusion coefficient would to be measured.

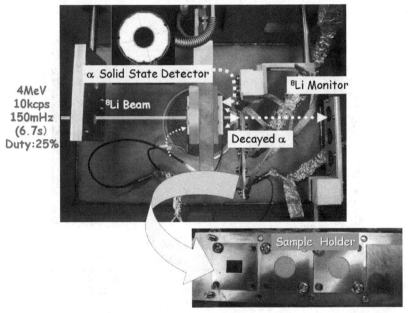

Fig. 3. Experimental set-up. The energetic and pulsed tracer beam of ⁸Li provided by the TRIAC are implanted into the sample and the decayed α particles are measured as a function of time by the solid state detector located in front of the sample. The condition of the tracer beam is given, which includes energy, intensity, repetition frequency of the pulsed beam and its duty factor for the beam-on time.

2.3 Data analysis for diffusion coefficients

Using the experimental set-up in Fig. 3, a test experiment has been performed to measure the diffusion coefficients in the sample of LiAl compound (Jeong et al., 2005a, 2005b). Indeed, the diffusion coefficient of Li has been successfully obtained with an accuracy of better than 25% as a result of the comparison between the experimental and simulated time-dependent yields of α-particles. In the following, we present how diffusion coefficients are extracted in our present method, by applying to the measurement of the Li diffusion coefficients in LiGa, which is an inter-metallic Li compound and known as a good Li ionic conductor.

Figure 4 shows a normalized time spectrum of the yield of α-particles measured at room temperature for Li_xGa_{1-x}, x=0.54 in atomic ratio. The spectrum is presented by the ratios (i.e. time-dependent yields of α-particles divided by the α radioactivity of ^8Li at the time of interest). In this way was excluded the trivial time-dependency in the yield of α-particles just governed by the lifetime of ^8Li. The values of the ratios, therefore, should be constant over time if ^8Li does not diffuse at all. However, the experimental values as shown in Fig. 4 gradually increase with time and then fall off. Based on the relative time-dependency of α-particle yields (i.e. time dependent ratios), a diffusion coefficient was extracted by comparing with a Monte Carlo simulation where one-dimensional Fickian (Gaussian) diffusion was assumed. In Fig. 4 is also presented the time spectrum simulated with the diffusion coefficient best reproducing the experimental data as a result of comparisons to be described in the following.

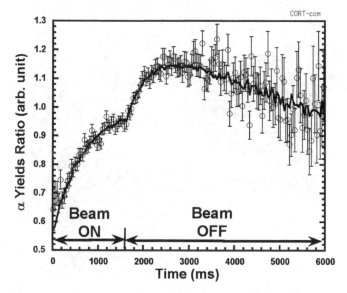

Fig. 4. Time spectrum of the α particle yields normalized by the α radioactivity of ^8Li implanted during the time of "Beam ON". The time spectrum, best simulated with the diffusion coefficient of $1.05 \times 10^{-7} cm^2/s$, is shown by solid line for comparison.

In the simulation, as described in detail in our previous publication (Jeong et al., 2003), we first defined the incident energy and energy spread of the ^8Li beam from the energy spectrum measured before implantation. Using the incident condition of the beam, we simulated the concentration-depth profile of ^8Li implanted in the sample and the time-evolution of the profile when a certain diffusion coefficient of Li in the sample was assumed. And then were simulated the energies of α-particles emitted from the time-dependent (diffusing) profiles of ^8Li, by taking into account the energy loss and straggling on their passage from the emitted position to the sample surface. Finally, integrated over the energies larger than 400 keV, the time-dependent α-particle yields associated with the diffusion coefficients assumed in the simulation were obtained and then compared with the experimental time spectrum, after being normalized in the same way as preformed for the experimental data. It should be noted that the resultant is the macroscopic diffusion coefficients, since the simulation neglects the isotopic characters of diffusing elements. The parameters (mean and FWHM) describing the concentration-depth profile, and the energy loss and straggling of α-particles were estimated by using the SRIM-2003 code (Ziegler, 1985), which is widely used in this kind of application with high reliability.

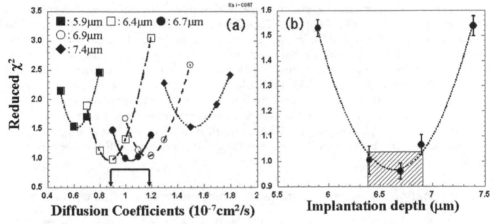

Fig. 5. (a) Reduced χ^2 values are compared for different implantation depths and correspondingly varied diffusion coefficients. Each of them was calculated with the spectrum simulated with a pair of implantation depth and diffusion coefficient. For each value of the implantation depth, the minimum value of the reduced χ^2 was extracted by a quadratic curve fitting as indicated by lines in (a), and then was plotted as a function of the corresponding depth in (b). By fitting the χ^2 values with a quadratic function of the depth [dotted line in (b)], the hatched region was estimated, especially as a constraint on the uncertain range of the implantation depth in the simulation. The allowed range of the depth was finally transformed into that of the diffusion coefficients [arrowed region in (a)], considering as a systematic error in the determination of the diffusion coefficient in the present case. (Fig. 2 from Jeong et al., 2008)

For comparison, we performed the χ^2 test, the likelihood test between the experimental and simulated time spectra. In the case of maximum likelihood, the reduced χ^2, the value of χ^2 simply divided by the number of data points in the present case, should be minimum and approximately close to 1. During the test of likelihood, two input parameters in the simulation were examined: diffusion coefficient and implantation depth. In the present simulation, the depth was explicitly considered as a variable input in order to take into account the roughness of the sample surface (less than 1μm) and range-uncertainty in the SRIM code. These two parameters were found to be most sensitive to the result of the simulation, i.e. the time-dependent structure of α-particle yields, but strongly inter-correlated in such a way that, e.g., simulations with somewhat higher (lower) diffusion coefficient and deeper (shallower) implantation could yield similar results as compared in Fig. 5(a). For different implantation depths, the χ^2 tests were explicitly performed with various diffusion coefficients. An example of the test is shown in Fig. 5(a), where the values of the reduced χ^2 were compared for different combinations of implantation depths and diffusion coefficients assumed in the simulation. For a value of the implantation depth, a diffusion coefficient giving rise to maximum likelihood (minimum value of the χ^2) can be identified. The minimum values of the χ^2 and the corresponding implantation depths are further compared in Fig. 5(b). Then a pair of the implantation depth and the diffusion coefficient corresponding to the most minimum value of the reduced χ^2 in the comparison is finally obtained; the diffusion coefficient of $(1.05\pm0.15)\times10^{-7}cm^2/s$ with the implantation depth of 6.64 (±0.25) μm in the present case. As indicated in Fig. 5 for the diffusion coefficients [arrowed in Fig. 5(a)] and implantation depth [hatched in Fig. 5(b)], the uncertainty in the determination of the coefficient comes mainly from the uncertain implantation depth. Here, we have considered the uncertain implantation depth as the main cause of systematic error inherent in the present method; otherwise the diffusion coefficient could be determined more accurately. It should be noted that the present results are not sensitive to the width of the primary concentration profiles of the tracer (Jeong et al., 2003).

As the reference spectrum for normalization, an experimental time spectrum of the α-radioactivity of ^8Li implanted in pure Cu was used. It allows us to avoid the systematic errors caused by the beam on/off operations, since no significant diffusion effects were observed in the case.

3. An application for measuring diffusion coefficients in Li ionic conductors

The β-phase of inter-metallic Li compounds, such as β-LiAl, LiGa and LiIn, has been considered as possible electrode materials in Li ionic batteries because of their high diffusion coefficients at room temperature for Li ions (Wen & Huggins, 1981). They are common in lattice structure; NaTl structure (Ehrenberg et al., 2002) composed of two interpenetrating sublattices, each forming a diamond lattice with a homogeneity range of around stoichiometric atomic ratio of Li (48~56 at. % Li for LiAl, 44~54% at. % Li for LiGa, 44~54 at. % Li for LiIn). In order to understand the motion of Li in the β phase as an ionic conductor, the defect structure in the Li compounds, closely related to the fast ionic motion, has been intensively studied via the measurement of electrical resistivity and density with help of the standard x-ray diffraction analysis (Sugai et al., 1995; Kuriyama et al., 1996). There exist three kinds of defects (see Fig. 6); vacancies on Li sites (V_{Li}), defects on anti-sites that replaced by Li (Li_A, A=Al, Ga, In) and complex defects (V_{Li} - Li_A). By forming the complex defects, the ionic motion of Li is suppressed or assisted depending on the kinds of anti-site atoms; Li diffusion rather slows down in β-LiAl while becomes rather faster in β-LiIn, although Li diffusivity almost linearly depends on the constitutional vacancy concentration on the Li sublattice (V_{Li}) (Tarczon et al., 1988). The high diffusion coefficient in the Li compounds is associated with the constitutional vacancy concentration on the Li sublattice, which is relatively large as compared to the usual metal alloy. The thermodynamic behavior of the Li vacancy has also been inferred from the anomalous electrical resistivity ("100K" anomaly) observed at around 95K near the critical composition corresponding to the Li-deficient region of β-LiAl (Kuriyama et al., 1980), which is considered as an order-disorder transition of vacancies on the Li sublattices (Brun et al., 1983).

The macroscopic ionic motion of Li has been so far inferred from the analysis of the electrical response to the applied voltage (electro-chemical method) (Sato et al., 1997). The values of diffusion coefficients, essentially obtained in such indirect ways, are often scattered over several orders of magnitude, strongly depending on the method of data analysis for finally extracting the diffusion coefficients.

Under such general situation, we have applied our method for measuring diffusion coefficients of Li, especially in β-LiGa where the diffusion coefficients has not been well measured although Li diffusion is known to be fastest among the Li inter-metallic compounds. The high diffusivity of Li in β-LiGa is associated with an especially large, constitutional vacancy concentration on the Li sublattice, almost three times lager than in β-LiAl. It would be also very interesting to observe, directly in terms of diffusion coefficient, the order-disordering transition of the vacancies on Li sites as well as the effect on the Li diffusion associated with the formation of the complex defects.

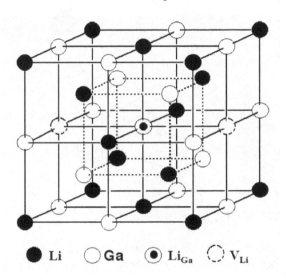

$\bullet$ Li $\bigcirc$ Ga $\odot$ Li$_{Ga}$ $\overset{\cdot}{\underset{\cdot}{\cup}}$ V$_{Li}$

Fig. 6. Crystal structure of NaTl-type inter-metallic compound LiGa contained defects (Li: Li atoms at the Li-sites, Ga: Ga atoms at the Ga-sites, Li$_{Ga}$: Li atoms at the Ga-sites (i.e. anti-site defects, where Ga-sites are replaced by Li atoms), V$_{Li}$: vacancies at the Li-sites). The most Li-poor β-LiGa (44 at. % Li) can have two V$_{Li}$ in unit cell, although the concentration of Li$_{Ga}$ is close to zero (see Fig. 7).

3.1 Materials
The samples of β-LiGa with the Li composition of 43~54 at. % were prepared by direct reaction of desired amounts of lithium (99.9%) and gallium (99.999%) in a tantalum crucible. The crystallization was performed by the Tammam-stöber method as reported in Yahagi, 1980. And the crystal was found to be polycrystalline by X-ray diffraction analysis.

The composition of Li was determined by the electrical resistivity measurements, relying on the systematic correlation between them (Kuriyama et al., 1996) as shown in Fig. 7. The concentrations of the point defects, [V$_{Li}$] and [Li$_{Ga}$], strongly depends on Li compositions; with increasing Li composition from 43 to 54 at. %, [V$_{Li}$] decreases from 11.4 to 2.8%, while [Li$_{Ga}$] increases from 0 to 5.1%. V$_{Li}$ is the dominant defect for Li-poor compositions, Li$_{Ga}$ is the dominant defect for the Li-rich ones, and mixing of the two defects extends throughout the entire phase region. The coexistence of V$_{Li}$ and Li$_{Ga}$ is expected to form V$_{Li}$-Li$_{Ga}$ complex defects (Kuriyama et al., 1996) as reported for the defect structure of β-LiAl (Sugai et al., 1995), which would play an important role in reducing the strain energy caused by the point defects in the lattice matrix. Especially, almost the same amounts of V_{Li} and Li_{Ga} exist for the composition of about 51 at. % Li.

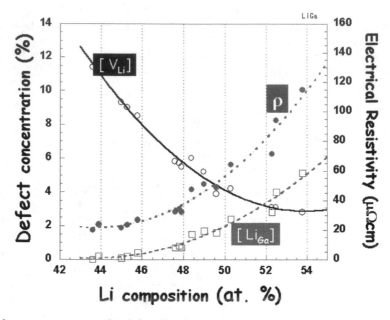

Fig. 7. Defect concentrations ($[V_{Li}]$, $[Li_{Ga}]$), electrical resistivity (ρ) vs. Li composition in atomic % for LiGa. The data points were taken from Kuriyama et al. 1996, and plotted for comparison.

The sample sliced in a form of disk with a diameter of 10mm and a thickness of 1mm was installed on the sample holder in the experimental chamber shown in Fig. 3. The surface of the sample was polished to the roughness less than 1 μm before set-up.

3.2 Li composition dependence of Li diffusion in Li inter-metallic compounds

Since the concentration of the defects is characterized by the Li composition under control in synthesis (Yahagi, 1980), as demonstrated in Fig. 7, the Li inter-metallic compounds have attracted much attention as a typical Li ionic conductor for investigating the high diffusivity of Li ions in a well-defined environment of defects. The detailed study on the diffusivity of Li in the Li inter-metallic compounds is further of interest, since the compounds have been considered as possible negative electrodes for Li ion secondary batteries more efficient than the commercially available (Wen & Huggins, 1981; Saint et al., 2005).

For β-LiAl and β-LiIn with the composition ranging between about 48 and 53 at. % Li, the diffusion coefficients of Li have been measured by a pulsed field gradient nuclear magnetic resonance (PFG-NMR) method in Tarczon et al., 1988, where was shown a strong correlation between the defect structure and the Li diffusion coefficient through their respective Li composition-dependency. The diffusion coefficient at room temperature was observed to become higher monotonously with decreasing Li composition - in other words, increasing concentration of V_{Li}. It was pointed out, furthermore, that the monotonous dependence of Li diffusivity on the Li composition was found to be slightly modified, depending on the species of the anti-site atoms.

While the monotonous composition-dependence of Li diffusivity in both β-LiAl and β-LiIn strongly suggested that the Li atoms diffuse via a vacancy mechanism (i.e. by a vacancy-atom exchange process through nearest neighbored paths on the Li sublattice), the slight modification in the correlation was partly understood by the coexistence of two types of defects, namely vacancies in the Li sublattice and Li anti-site atoms in the aluminum or indium sublattice forming compound defects (V_{Li}-Li_{Al}, or V_{Li}-Li_{In}) (Kishio & Britain, 1979; Tarczon et al., 1988). Owing to the elastic relaxation of atoms against the strain induced in the vicinity of the point defects, the interaction between the two types of defects once forming a compound defect can be attractive or repulsive according to the relative size of the ions on the sublattice replaced by Li (atomic size effect). The Li anti-site atom Li_{Al} in β-LiAl produces compressional strain (expanded lattice), since the radius (0.68Å) for the Li ion in a closed shell configuration (Kittel, 2005) is larger than that (0.50 Å) for the aluminum ion, while the anti-site atom Li_{In} in β-LiIn induces dilatational strain (contracted lattice) because of ionic radius (0.8 Å) for indium larger than that of the substitutional Li ion. On the other hand, the vacancy V_{Li} always produces dilatational strain. Indeed, as supposed by the atomic size effect, the diffusion in β-LiIn was observed to be enhanced by the presence of the repulsive interaction between different types of defects, more than expected by the single vacancy diffusion mechanism (Tarczon et al., 1988).

Based on the atomic size effect, the interaction between V_{Li} and Li_{Ga} in β-LiGa is supposed to be attractive as in β-LiAl, because the radius (0.62 Å) of gallium ion is slightly smaller than that of Li ion. The strength of the interaction in β-LiGa is expected to be weaker than observed in β-LiAl and β-LiIn, since the radii of the constituent ions are quite close to each other. In addition to the specific interaction – attractive and weak, LiGa has stable β-phase over the Li composition range wider than investigated so far in β-LiAl and β-LiIn, consequently allowing us to address the Li diffusion in the concentration range of V_{Li} almost three times wider than before (Kuriyama et al., 1996).

The diffusion coefficients of Li in β-LiGa have not yet been measured in detail, but measured only by the electro-chemical method (Wen & Huggins, 1981), where the Li composition was not well defined. Using the on-line diffusion tracing method introduced in the previous section 2, we have measured the diffusion coefficients of Li in β-LiGa with the composition in the range of about 43 to 54 at. % Li. Of special interest is how the Li diffusion in β-LiGa depends on the Li composition, consequently the concentration of the V_{Li} defect.

3.2.1 Abnormal Li diffusion: Enhanced or suppressed by the formation of defect complex

In Fig. 8, the time-dependent normalized α-particle yields are compared for different Li compositions, i.e. 43.6, 50.0 and 53.2 at. % Li. Referring to the time dependence of the α-particle yields, the stoichiometric β-LiGa has the highest diffusivity of Li among three, most quickly rising and falling down. This observation is quite different from those observed for β-LiAl and β-LiIn (Tarczon et al., 1988) which are iso-structural with the β-LiGa.

The abnormal behavior of Li diffusivity in β-LiGa is well identified in Fig. 9, where the diffusion coefficients in β-LiGa at room temperature are presented as a function of Li composition together with the data for β-LiAl and β-LiIn reported in Tarczon et al., 1988. As shown in Fig. 9, the Li diffusion coefficients in β-LiAl and β-LiIn decrease monotonously with increasing Li composition, but the correlation is changing due to the coexistence of vacant Li sites V_{Li} and anti-site Li atoms on the aluminum (or indium) sites Li_{Al} (or Li_{In}), as discussed earlier in terms of the size effect of constituent atoms. Assuming the single vacancy diffusion in LiGa, the diffusion coefficients are supposed to keep increasing in the Li-poor composition in similar ways as observed in both β-LiAl and β-LiIn. However, what observed in the present measurement goes the other way; the diffusion becomes faster around stoichiometric atomic ratio, i.e. showing a maximum around the Li composition of 48~50 at. %.

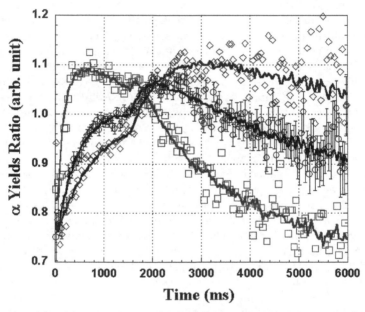

Fig. 8. Normalized time spectra of α yields obtained at room temperature for β-LiGa with Li atomic compositions of 43.6(○), 50.0(□) and 53.2(◇) at. %. The results from the simulation are also shown as solid lines. In the simulation, best reproducing the data as a result of χ^2 tests, the diffusion coefficients of $2.4 \times 10^{-7} cm^2/s$, $8.5 \times 10^{-7} cm^2/s$ and $1.2 \times 10^{-7} cm^2/s$ were assumed for 43.6, 50.0 and 53.2 at. % Li, respectively.

Although the data for β-LiAl and β-LiIn were available only in a limited range of Li composition from 48.3 to 53 at. % Li, it would be interesting to note that the diffusion coefficients in β-LiGa varied in a similar way (increase with decreasing Li content) between those values in β-LiAl and β-LiIn in the corresponding range of Li composition. This would be intuitively understandable, since the iso-structural β-LiGa with a comparable size of the constituent atoms should have an interaction between V_{Li} and Li_{Ga} with intermediate strength as compared to those discussed in the case of β-LiAl and β-LiIn. Therefore, the composition-dependence of Li diffusivity in β-LiAl and β-LiIn could be considered as the lower and upper limits of the Li diffusivity in β-LiGa over the Li composition range of interest, respectively; two complementary explanations might be possible by referring to the respective composition-dependency observed in β-LiAl and β-LiIn.

Referring to the tendency in β-LiIn, the diffusion of Li in very Li-poor β-LiGa seems to be suppressed. This could happen, for example, by assuming the formation of defect complex such as V_{Li}-V_{Li} and/or V_{Li}-Li_{Ga}-V_{Li}, since the concentration of V_{Li} become much (almost three times at maximum) larger (Kuriyama et al., 1996) in the more Li-poor composition than investigated in β-LiIn (Tarczon et al., 1988). It should be noted that the number of vacant Li sites in a unit cell volume (8 for Li and 8 for Ga) is about two for the most Li-poor β-LiGa (43.6 at. % Li), whereas there exist about one vacant Li site in every two-unit cells for the most Li-poor β-LiAl (48.3 at. % Li) and β-LiIn (48.4 at. % Li) (refer to Fig. 6). Here, we assumed the random distribution of the vacancies over the available sites.

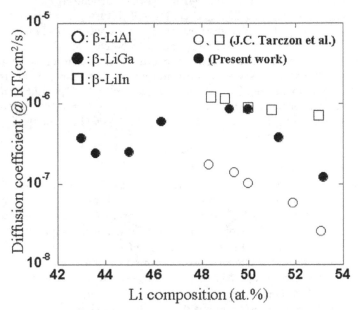

Fig. 9. Li composition-dependence of diffusion coefficients of Li at room temperature. The diffusion coefficients measured at room temperature are given as function of Li composition in atomic % for β-LiGa (present work), β-LiIn and β-LiAl (from Tarczon et al., 1988), respectively. The most Li-poor composition of LiGa shown in the figure is out of the β-phase (Fig.5 from Jeong at al., 2009).

Alternatively, assuming the diffusivity of Li in the most Li-poor β-LiGa as a good extension of the diffusivity of Li in the Li-poor β-LiAl (48.3 at. % Li) because of nearly zero concentrations of Li_{Al} and Li_{Ga} in the corresponding Li composition, the diffusivity of Li observed around the stoichiometric β-LiGa could be considered to be enhanced by the coexisting defects of V_{Li} and Li_{Ga}. Under the coexistence of V_{Li} and Li_{Ga}, the motion of the vacancies on the Li site, supposedly the carriers of Li atom, seems to be strongly promoted. This suggests that the interaction between V_{Li} and Li_{Ga} would be stronger and rather repulsive than expected by the atomic size effect, but not as strong as observed in β-LiIn. In addition, as a specific characteristic of the interaction in β-LiGa, we found strong composition-dependence of the interaction that appears to become stronger when a comparable amount of two types of point defects exists.

In the present measurement, although we found abnormal Li diffusion in very Li-poor composition of β-LiGa, the characteristic of the anomaly in diffusion, i.e. whether the diffusion is suppressed or enhanced, is not conclusive. A detailed theoretical consideration is highly required for quantitative discussion. From the experimental point of view, however, it would be interesting to extend the measurement to the more Li-poor composition of LiIn recently confirmed to have the β-phase in the same composition (Asano, 2000) as investigated for β-LiGa in the present work.

So far, we have compared our data for LiGa with those obtained by a pulsed field gradient nuclear magnetic resonance (PFG-NMR) method in Tarczon et al, 1988. The chemical diffusion coefficients of Li in LiGa obtained by an electrochemical method at 415°C were reported earlier (Wen & Huggins, 1981), where the self-diffusion coefficients of Li were also calculated by considering that the diffusivity of Ga was appreciably lower that that of Li. The self-diffusion coefficients at 415°C were found to be constant 1.1×10^{-6} and 5.0×10^{-7} cm^2/s on the Li poor and rich sides of LiGa, respectively, with a sudden change between 48 and 47.6 at. % Li. Those values were almost one order of magnitude smaller than those extrapolated to the temperature of 415°C from our present data. It should be noted that the similar amount of discrepancies in the values of diffusion coefficients were also found in the case of LiAl, by comparing the data obtained by PFG-NMR (Tarczon et al, 1988) and the electrochemical method (Wen et al, 1979).

3.3 Order-disordering of Li vacancies

Figure 10 shows time spectra of the yield of α-particles (represented by the ratios normalized to 1 at the end of beam-on) measured at different temperature for the stoichiometric LiIn. With decreasing temperature, the more α-particles are observed at later time, which means that the diffusion becomes slower at lower temperature. Interestingly, there observed a big change in the time structure of α-particle yields (normalized) at a certain temperature, where the temperature was varied by almost the same amount for the neighboring measurements compared in Fig. 10. Based on this relative time-dependency of α-particle yields (time dependent ratio) as demonstrated in Fig. 10, diffusion coefficients at various temperatures could be estimated by the way discussed previously.

In order to observe such a sudden change for β-LiGa, we have also performed a detailed measurement, where the temperature varied by a fine step (about 5-degree difference in temperature between neighboring measurements), especially below room temperature.

The diffusion coefficients of Li in β-LiGa and β-LiIn with a near stoichiometeric composition of Li are displayed in Fig. 11 as a function of inverse temperature. For both samples, the diffusion coefficients suddenly change at a certain temperature, as observed time-dependent ratio spectra in Fig. 10, and follow Arrhenius behavior in the region of higher temperature. The sudden change in the value of the diffusion coefficient for β-LiGa of 44 at. % Li occurs at 234 (±2) K. The anomalous electrical resistivity (i.e. sudden change in the value of resistivity) is also observed at the same temperature, as shown in Fig. 11. The resistivity measurements were carried out using a van der Pauw method as used for β-LiAl (Sugai et al., 1995). This observation is closely related to the thermal properties of the structural defects, already observed as the anomalies in heat capacity (Hamanaka et al., 1998; Kuriyama et al., 1986) and nuclear-spin lattice relaxation (Nakamura et al., 2007) at 233 K near the critical composition of the Li-poor β-LiGa. It has been suggested that these phenomena are related to order-disorder transformation of the Li vacancies in the compounds. By the neutron diffraction measurements for β-LiAl (Brun at al., 1983), a sudden change around 100K in the electrical resistance ("100K" anomaly as mentioned previously) has been understood by an ordering of the vacancies on the Li sublattice below the transition temperature; the structure below 100K is body centered tetragonal; the unit cell contains 10 Li and 20 Al positions, with the vacancies located at 2a positions.

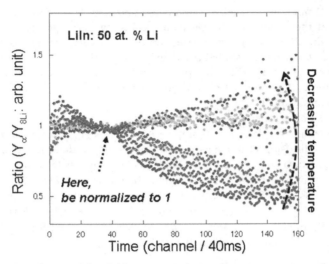

Fig. 10. Time spectra of α-particle yields measured at various temperatures for LiIn. The spectra were corrected for removing trivial time dependence and further normalized properly for easy comparison.

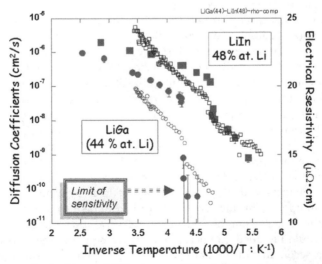

Fig. 11. Temperature-dependence of diffusion coefficients (closed symbols) and electrical resistivity (open symbols) for β-LiGa with 44 at. % Li, and β-LiIn with 48 % at. Li.

The ordering of the vacancies would produce a sharp drop in the Li diffusion coefficients at the ordering temperature, since the vacancies are supposed to be carriers of Li atom (i.e. ordered vacancies suppress the random exchange of Li atoms via vacancies). The observed amount of change, more than two orders of magnitude in the value of diffusion coefficients in the case of β-LiGa, is quite impressive as compared to those observed in the measurement of electrical resistivities where just a small change (at most 1/10) can be seen at the transformation temperature. The transition temperature is known to shift to lower temperature, and the change in the electrical resistivity becomes invisibly smaller with increasing Li composition, where a large fraction of the Li vacancies forms complex defects with anti-site Li atoms on Ga sublattices (Kuriyama et al, 1986). Therefore, with higher sensitivity, the present method could be applied for better investigating the characteristics of the transformation that are supposed to be correlated with the concentrations of structural defects (i.e. Li compositions).

Around the transition temperature, the time structure of the normalized α-particle yields could not be explained in terms of one component of diffusion, i.e. one diffusion coefficient, implying that the diffusion coefficients would not be singly determined in the transition region. The temperature window, in which the transition appears to occur, is about 10K.

At the lower temperature followed by a sudden change around 234 (±2) K for β-LiGa, the diffusion coefficients are observed as a constant, which is the lower limit of diffusion coefficients accessible by the present method; for diffusion coefficients less than about 10^{-10} cm^2/s, any significant effect in the yields of α-particles due to the diffusing ^{8}Li could not be observed because of the short life-time of the radiotracer.

4. Aiming at lithium micro- and nano- scope

As an extension of the present method, we consider the possibility to trace the Li macroscopic behavior across the interface in hetero-structural Li ionic conductors in micrometer scale, to be termed ^{8}Li microscope.

The time spectra shown in Fig. 10 represent dynamical movement of Li in the sample between as-implanted position and surface during a cycle for measurement. For a single layer (diffusion in homogeneous sample) as discussed in case of Fig.2, the present method can trace, most efficiently, the Li movement within one-dimensional distance of about 7μm for about 7s. For slow diffusion, Li is still moving toward the surface for the time of measurement, the α-particle yields are monotonically increasing with time. For moderate diffusivity, a maximum is observed when Li is reflected by the surface. After the maximum, the yield is simply decreasing with time since Li is diffusing into the bulk. For double layers (hetero-structural sample) whose interface exist in between, i.e. introducing an interface between the as-implanted position and the surface, we could observe time structure different from the case of single layer. In other words, we could observe how Li interacts with the interface, e.g. if Li is precipitated, perfect- or half-reflected on the interface. This idea could be applied to take a dynamical picture of Li in Li ion micro-batteries consisting of thin films of several μm in thickness. Therefore, the present method could be called ^{8}Li microscope by analogy with the neutron transmission image of Li in a secondary Li ion battery, where the picture of Li, actually ^{6}Li, in the battery was taken by the neutron radiography with a resolution of mm (Takai et al., 2005).

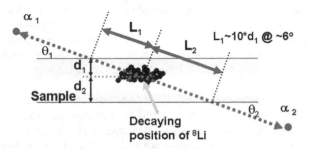

Fig. 12. Schematic layout for a coincident measurement of two α-particles emitted with angles of $\theta_1=\theta_2$ relative to the surface of sample. The actual path lengths (L_1 and L_2) in the sample are enhanced 10 times as compared to those considered in Fig. 2 when applying limitation to the emission angle of 6°.

For higher sensitivity, from micro-scale to nano-scale, we consider a coincident measurement of two α-particles emitted from position of β-decaying ^{8}Li. The micrometer sensitivity of the method discussed so far is partly coming from the broad energy distribution of α-particles on decaying. Therefore, the coincident measurement is supposed to dramatically improve the sensitivity since the coincident α-particles have the same energy at the decaying position. In diffusion tracing method by ^{8}Li, however, a special attention is paid on the energy loss of α-particles subjected to the actual path length in the sample of interest from the decaying to the detection positions. For diffusion in a nano-scale, the diffusion length is too short to give a significant change in the energy loss of α-particles. We further apply a limitation in emission (detection) angles to the coincident measurement as shown in Fig. 12. With small emission angles, the actual path length of α-particles in the sample of interest can be made significantly longer than implanted depth, giving rise to a considerable energy loss difference against a nano-scale diffusion length toward the surface.

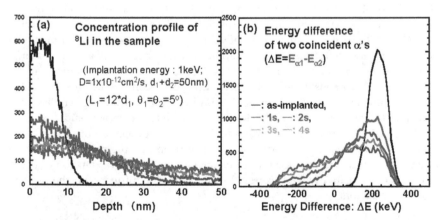

Fig. 13. (a) Simulated concentration profiles of ^{8}Li implanted into $LiCoO_2$ with a thickness of 50nm with an energy of 1keV and their evolution by diffusion simulated at 1, 2, 3 and 4 seconds after implantation with a diffusion coefficient of 10^{-12} cm^2/s. (b) Spectra of energy difference of α-particles coincidently emitted at angles of 5° relative to the surface of the sample. They are respectively simulated in the same time sequence as done in Fig.13 (a).

Along with the idea, we have performed a simulation to examine the feasibility and search for an observable most sensitive to diffusion profiles with a diffusion coefficient of 10^{-12} cm²/s, a specific goal of present consideration. The simulation was performed in a similar way discussed in Jeong et al, 2003, by using the energy loss and straggling of α-particles provided by the SRIM-code. We assumed the diffusion coefficient as 10^{-12} cm²/s, the sample thickness as 50nm, the emission angle as 5º, and the implantation energy as 1keV.

As a result of the simulation shown in Fig.13, we found that the energy difference between two coincidently measured α-particles could provide nano-scale sensitivity. The concentration profiles of ^{8}Li simulated sequentially with a condition given above are shown in Fig. 13 (a); the as-implanted profile is broadening with time by diffusion. Simulated in the same time sequence as done for profiles, the spectra of energy difference of two α-particles to be measured coincidently are compared in Fig. 13 (b). There is a good correspondence between the diffusion profiles and energy difference spectra, more specifically a clear one-to-one correspondence between decaying position of ^{8}Li and energy difference of two coincidently measured α-particles. For example, looking at the time evolution of the counts of coincident events with zero energy difference in the energy difference spectra should be a good measure of the time when ^{8}Li is across the middle of the sample, because the zero energy difference means that the path lengths experienced by two coincident α-particles are identical. We can conclude that the sensitivity against the diffusion of 10^{-12} cm²/s could be easily achieved by the coincident measurement of two α-particles with the geometry assumed in the simulation.

5. Conclusions

A non-destructive and on-line radiotracer method for diffusion studies in lithium ionic conductors has been reviewed. As the tracer, the pulsed beam of the short-lived α-emitting radioisotope of ^{8}Li was implanted into a sample of interest. By analyzing the time-dependent yields of the α-particles from the diffusing ^{8}Li measured in coincident with the repetition cycle of the beam, the tracer diffusion coefficients were extracted with a good accuracy. The method has been successfully applied to measure the lithium diffusion coefficients in a typical defect-mediated lithium ionic conductor of LiGa, well demonstrating that the method is very efficient to measure the diffusion in the micro-meter regime per second. Anomalous composition-dependence of Li diffusion coefficients in β-LiGa was observed; the stoichiometric LiGa showed the highest diffusivity of Li. The anomaly was discussed qualitatively in terms of the formation of defect complex and the interaction between the constituent defects. The ordering of the Li vacancies in the Li-deficient LiGa was observed for the first time in terms of the Li diffusion by the present method, and its thermodynamic aspect was discussed. Further development, as an extension of the present method for higher sensitivity, was proposed to measure the diffusion on the nano-scale in lithium ionic conductors.

6. Acknowledgments

The author would like to thank all the members of the TRIAC collaboration between KEK (I. Katayama, H. Kawakami, Y. Hirayama, N. Imai, H. Ishiyama, H. Miyatake, Y.X. Watanabe, S. Arai, K. Niki, M. Okada, M. Oyaizu), JAEA (A. Osa, Y. Otokawa, M. Sataka, H. Sugai, S.

Ichikawa, S. Okayasu, K. Nishio, S. Mitsuoka, T. Nakanoya), and Aomori University (M. Yahagi, T. Hashimoto). The present studies have been greatly indebted to the staffs of the Tandem Accelerator Center at JAEA for ensuring stable delivery of primary beams for producing radiotracer beam of ^{8}Li at the TRIAC. The work is partly supported by a Grant-in-Aid for Scientific Research (B), No. 16360317 from the Japan Society for the Promotion of Science.

7. References

Bonner, P. W. et al. (1948). A study of breakup of ^{8}Li, *Physical Review*. Vol. 73, Issue 8, (April 1949), pp. 885-890, 0031-899X

Brun, T. O. et al. (1983). Ordering of Vacancies in LiAl, *Solid State Communications*, Vol. 45, Issue 8, (Feburary1983), pp. 721-724, 0038-1098

Chandra, S. (1981). *super-ionic solids: principle and applications*, North-Holland Publishing Company, ISBN 0-444-86039-8, New York

Cornell, J. (Ed.). (December 2003). Solid-Sate Physics at EURISOL, In: *The Physics Case for EURISOL*, GANIL, available from http://pro.ganil-spiral2.eu/eurisol/feasibility-study-reports/feasibility-study-appendix-a

Hamanaka, H. et al. (1998). Anomalous heat capacity and defect structure in β-LiGa, *Solid State Ionics*, Vol. 113-115, (December 1998), pp. 69-72, 0167-2738

Ichikawa, S. (2003). Ion source development for the JAERI on-line isotope separator, *Nuclear instruments & methods in physics research B*, Vol. 204, (*May 2003*), pp. 372-376, 0168-583X

Ehrenberg, H. et al. (2002). Phase Transition from the Cubic Zintl Phase LiIn into a Tetragonal Structure at Low Temperature, *Journal of solid state chemistry*, Vol. 167, Issue 1, (August 2002), pp. 1-6, 0022-4596

Jeong, S. C. et al. (2003). Simulation Study on the Measurements of Diffusion Coefficients in Solid Materials by Short-lived Radiotracer Beams, *Japanese journal of applied physics*, Vol. 42, No. 7A, (July 2003), pp. 4576-4583, 0021-4922

Jeong, S. C. et al, (2005a). Measurement of diffusion coefficients in solids by the short-lived radioactive beam of ^{8}Li, *Nuclear Instruments and Methods in Physics Research B*, Vol. 230, Issues 1-4, (April 2005), pp. 596-600, 0168-583X

Jeong, S.C. et al. (2005b) Measurement of self-diffusion coefficients in Li ionic conductors by using the short-lived radiotracer of 8Li, *Journal of Phase Equilibria and Diffusion*, Vol. 26, No. 5, (September 2005), pp. 472-476, 1547-7037

Jeong, S.C. et al. (2008) On-Line Diffusion Tracing in Li Ionic Conductors by the Short-Lived radioactive Beams of 8Li, *Japanese journal of applied physics*, Vol. 47, No. 8, (July 2008), pp. 6413-6415, 0021-4922

Jeong, S.C. et al. (2009). Abnormal Li diffusion in b-LiGa by the formation of defect complex, *Solid State Ionics*, Vol. 180, Issues 6-8, (May 2009), pp. 626-630, 0167-2738

Kishio, K & Britain, J.O. (1979). Defect structure of β-LiAl, *Journal of Physics and Chemistry of Solids*, Vol. 40, Issue 12, (1979), pp. 933-940, 0022-3697

Kittel, C. (2005) *Introduction to Solid State Physics* (8-th ed.), Wiely, New York, p71

Kuriyama, K. et al. (1996). Defect Structure and Li-Vacancy Ordering in β-LiGa, *Physical review. B; Condensed matter and materials physics*, Vol. 54, Issue 9, (September 1996), pp. 6015-6018, 1098-0121

Kuriyama, K. et al. (1986). Electrical-transport properties in the semimetallic compound LiGa, *Physical review. B; Condensed matter and materials physics*, Vol. 33, Issue 10, (May 1986), pp. 7291-7293, 1098-0121

Kuriyama, K., Kamijoh, K & Nozaki, T. (1980). Anomalous Electrical Resistivity in LiAl Near Critical Composition, *Physical review. B; Condensed matter and materials physics*, Vol. 22, Issue 1, (July 1980), pp. 470-471, 1098-0121

Prandolini, M. J. (2006). Magnetic nanostructure: radioactive probes and recent developments, *Reports on Progress in Physics*, Vol.69, No. 5, (May 2006), pp. 1235-1324, 0034-4885

Nakamura K. et al. (2007). Li+ ionic diffusion and vacancy ordering in β-LiGa, *Faraday Discussions*, Vol. 134, (2007), pp. 343-352, DOI: 10.1039/B602445A

Saint, J. et al. (2005). Exploring the Li-Ga room temperature phase diagram and the electrochemical performance of the Li_xGa_y alloys vs. Li, *Solid State Ionics*, Vol. 176, issues 1-2, (January 2005), pp. 189-197, 0167-2738

Sato, H. et al, (1997). Electrochemical characterization of thin-film $LiCoO_2$ electrodes in propylene carbonate solutions, *Journal of Power Sources*, Vol. 68, Issue 2, (October 1997), pp. 540-544, 0378-7753

Sugai, H. et al. (1995) Defect Structure in Neutron-Irradiated b-6LiAl and b-7LiAl: Electrical Resistivity and Li Diffusion, *Physical review. B; Condensed matter and materials physics*, Vol. 52, Issue 6, (August 1995), pp. 4050-4059, 1098-0121

Takai S. et al. (2005). Diffusion coefficients measurements of La2/3-xLi3xTiO3 using neutron radiography, *Solid States Ionics*, Vol. 176, Issues 39-40, (September 2005), pp. 2227-2233, 0167-2738

Tarczon, J. C. et al. (1988).Vacancy-Antistructure Defect Interaction Diffusion in β-LiAl and β-LiIn, *Materials Science &. Engineering A*, Vol. 101, (May 1988), pp. 99-108, 0921-5093

Tuijn, C. (1997). On the history of Solid-state diffusion. *Defect and Diffusion Forum*, Vol. 141-142, (1997), pp. 1-48, 1662-9507

Yahagi, M. (1980). Single crystal growth of LiAl, *Journal of Crystal Growth*, Vol. 49, Issue 2, (June 1980), pp. 396-398, 0022-0248

Watanabe, Y. X. et al. (2007). Tokai Radioactive Ion Accelerator Complex (TRIAC), *European Physical Journal - Special Topics*, Vol. 150, No. 1, (March 2007), pp. 259-262, 1951-6355

Wen, C. J. & Huggins, R. A. (1981). Electrochemical Investigation of the Lithium-Gallium System, *Journal of The Electrochemical Society*, Vol. 128, issue 8, (August 1981), pp. 1636-1641, 0013-4651

Wen, C. J. et al. (1979). Thermodynamic and Mass Transport properties of "LiAl", *Journal of The Electrochemical Society*, Vol. 126, Issue 12, (December 1979), pp. 2258-2266, 0013-4651

Wenwer, F. et al. (1996). A universal ion-beam-sputtering device for diffusion studies, *Measurement science & technology*. Vol. 7, No. 4, (April 1996), pp. 632-640, 0957-0233

Wichert, T. & Deicher, M. (2001). Studies of semiconductors, *Nuclear Physics A*, Vol. 693, No. 3-4, (October 2001), pp.327-357, 0375-9474

Ziegler, J. F., Biersack, J. F. & Littmark, U. (1985). *The Stopping and Range of Ions in Solids* (2003 Version), Pergamon Press, New York, Chap. 8, available from http://www.srim.org/

3

Research Reactor Fuel Fabrication to Produce Radioisotopes

A. M. Saliba-Silva[1], E. F. Urano de Carvalho[1],
H. G. Riella[2] and M. Durazzo[1]

[1]Nuclear Fuel Center of Nuclear and Energy Research Institute
Brazilian Commission of Nuclear Energy, São Paulo,
[2]Chemical Engineering Department of University of Santa Catarina, Florianopólis,
Brazil

1. Introduction

This chapter describes the manufacturing technology of fuel used in research reactors that produce radioisotopes. Besides this production, the research reactors are also used for materials testing. The most common type of research reactors is called "MTR" - Materials Testing Reactor. The MTR fuel elements use fuel plates, which are quite common around the world. There was a historic development in that fuel type over the years to reach the current state-of-art in this technology.

The basic MTR fuel element is an assembled set of aluminum fuel plates. It consists of regularly spaced plates forming a fuel assembly. These spaces allow a stream flow of water that serves as coolant and also as moderator to nuclear reaction. The fuel plates have a meat containing the fissile material, which is entirely covered with aluminum. They are manufactured by adopting the traditional assembling technique of dispersion fuel briquette inserted in a frame covered by aluminum plates, which are welded with subsequent rolling. This technique is known internationally under the name "picture-frame technique". Powder metallurgy techniques are used in the manufacture of the fuel plate meats, making briquettes using ceramic or metallic composites. The briquette is made with powdered nuclear material and pure aluminum powder, which is the structural material matrix of the briquette.

Using UF_6 in the chemical plant, it is able to produce several intermediate compounds of uranium. One of these compounds is UF_4, which is the main raw material to produce metallic uranium. It could be made by several routes. The production of metallic uranium uses the UF_4 reduction through calcio- and magnesiothermic reaction. The metallic uranium is alloyed with Al, Si or Mo. Previously, stable uranium oxides were used as MTR fuels, but they had very small densities to accomplish a good operational performance of the reactors. The fuel material candidate mostly prone to be used in nuclear research reactors is based on alloys carrying more U-density toward the fuel meat. In present state, the U-Mo alloys are good candidates, but it would not be subject of the present chapter since it on its path to be certified to future use in research reactors. Currently, the most used material is U_3Si_2 LEU, which is low enriched uranium enriched up to 20% of ^{235}U isotope, which is the nuclear fissile material.

The production procedures of U_3Si_2 fuel fabrication will be discussed in this chapter, starting from U_3Si_2 fabrication and powder manufacture. This powder is mixed with aluminum powder and pressed, resulting in a solid briquette with good mechanical strength. After quality inspection, the briquette becomes the fuel plate meat.

The fuel plate manufacturing procedures will be described according to the picture frame technique. This technique includes the assembling of the briquette inside the frame sandwiched with cover plates. The assembly is welded and hot and cold rolled to get the fuel plate, where the fuel meat is completely sealed inside aluminum. All the process and quality control during fabrication will be commented ahead.

Once the plates having been fabricated, the fuel assembly is finally made by fixing the fuel plates and the other mechanical components, such as nozzle, handle and screws. This finishing process to produce the element is also commented in this chapter.

The characteristics of the fuel plates must meet specifications of each particular research reactor characteristics. Inspections and qualifications are carried out in various stages of fuel plate manufacture.

As this chapter describes nuclear fuel manufacturing for research reactors, the sub-items of fabrication process can be divided into the following topics: evolution of nuclear fuel materials for MTR fuel; production of uranium hexafluoride (UF_6); production of uranium tetrafluoride (UF_4); production of metallic uranium; U_3Si_2 production; production of fuel cores from U_3Si_2 powder and aluminum; production of fuel plates with U_3Si_2-Al dispersion briquettes; assembling of fuel elements; recovery of uranium; effluent treatment; quality control.

In this chapter, the experience of IPEN/CNEN-SP (Energy and Nuclear Research Institute of Brazilian Commission of Nuclear Energy, São Paulo, Brazil) will be given as a productive route to produce MTR nuclear fuels for research reactors, since this is the main expertise of all the authors of this chapter.

2. Evolution of nuclear materials for research reactors fuel

The use of radioisotopes in medicine is certainly one of the most important social uses of nuclear energy. Radiopharmaceuticals are radioactive substances that help doctors to make important decisions for treatments in oncology, cardiology, neurology, among other areas. For patients, the diagnoses represent safety and pain relief, as in the case of samarium-153 use, which is employed to relieve bone pain caused by metastatic tumors.

Nuclear medicine is a medical specialty that uses radioactive material for diagnostic tests and therapeutic purposes. Although it is often confused with radiotherapy, the last application has a lot of different procedures and applications. The main distinction between the two specialties is the way both use the radioactive material. While radiotherapy (radiation therapy) uses sealed sources (or closed), which emit radiation outside the patient, nuclear medicine uses open sources of radiation, administered in vivo (oral or intravenous). If, in radiotherapy, radiation is directed toward the point to be discussed, in nuclear medicine is the body own metabolism of the patient who is in charge of carrying radioactive material into the organ to be examined or treated.

The success of nuclear medicine in diagnosis is due to its ability to show the functioning of various body organs, avoiding the use of invasive techniques such as biopsy and catheterization. The use of ultrapure iodine-123 to examine thyroid function is one example. By scintigraphy, a diagnostic imaging technique, which has several medical applications, made possible to measure the uptake of iodine by the thyroid and thereby assess the functioning of the gland. Another radioactive element widely used to study the various functions of the human organism is technetium-99m. This isotope can be chemically combined with various organic complexes, which evaluate liver disorders, bone and brain, among others. In bone scintigraphy, the radioactivity of technetium reveals the existence of tumors from six to eight months before they have reached sufficient size to be picked up by X-ray examinations. With this, it is possible to start treatment much earlier with greater cure perspective.

Nuclear reactors that produce radioisotopes are called research reactors. This type of reactor is also used to perform tests on materials and nuclear fuels in the development phase. The modern research reactors are designed with both purposes, radioisotope production and testing of materials, and for this reason are called Multipurpose Reactors.

Unlike power reactors, which are well known and are intended to generate heat for electricity generation, the research reactors or the modern multipurpose reactors aim to generate neutrons used for radioisotopes production or for testing materials in terms of verify their performance under irradiation. Unlike power reactors, research reactors operating with much higher power density, which is necessary to get high neutron fluxes. For this reason its fuel is usually in the form of a metal plate, usually covered by aluminum. They are very different from the fuel rod with ceramic pellets (UO_2) as used in the fuel for power reactors.

The research reactors moderated and cooled with light water and using plate-type fuel elements has been named MTR type reactors (Materials Testing Reactor). After the construction of the first MTR, a joint venture of ORNL (Oak Ridge National Laboratory) and ANL (Argonne National Laboratory) operated it since March 31, 1952. Many research reactors around the world uses MTR type fuel elements, which are formed by assembling fuel plates fabricated by a well-known and established technique of assembling a core, commonly named fuel meat, which incorporates the fissile material, a frame plate and two cladding plates, with subsequent deformation by hot and cold-rolling (picture frame technique) (1) (2).

Initially, the fuel plates usually used as the core material an uranium-aluminum alloy (U-Al) containing 18 wt% of highly enriched uranium (93 wt% [235]U) (1) (3). Even in the 50's, with the concern about nuclear weapons non-proliferation, the research reactors began to use fuels containing low-enriched uranium (20 wt% [235]U) (4). With enrichment lowering, in order to maintain the reactivity and lifetime of the reactor cores, it became necessary to increase the amount of uranium in each fuel plate. In the U-Al alloy, the uranium concentration had to be increased to 45 wt% to compensate the decrease in the enrichment level.

Fuel plates containing the meat based on the U-Al alloy with 18 wt% of highly enriched uranium were easily fabricated. However, difficulties arise in fabricating fuel plates with meats of U-Al alloy containing 45 wt% of low-enriched uranium, because of the fragility and propensity for segregation of this alloy (4) (5) (6). An alternative to overcome this problem was the use of cores manufactured by powder metallurgy, which used dispersions of uranium compounds in aluminum and could incorporate quantities of low-enriched uranium significantly greater. For instance, the Argonauta reactor (10 MW), in Rio de Janeiro, Brazil, started its operation in 1956 and was developed by the Argonne National Laboratory, USA. This pioneer Brazilian research reactor used fuel plates with the meat based on an U_3O_8-Al dispersion containing 39 wt% of U_3O_8 with low enrichment (7).

Efforts were made to increase the concentration of uranium in this type of dispersion fuel, getting 65 wt% of U_3O_8 in the fuel fabricated for the Puerto Rico Research Reactor of the Puerto Rico Nuclear Center to the end of the 70's (8).

Aiming at obtaining more and more high neutron fluxes, the development of research reactors with higher power required a continuous production of fuels, which used highly enriched uranium (93 wt% ^{235}U), yielding higher specific reactivity and economics, since these fuels could stay longer in the reactor core (long life). The 100 MW HFIR (High Flux Isotope Reactor) used dispersion U_3O_8-Al with 40 wt% U_3O_8 (9) and the ATR (Advanced Test Reactor), with 250 MW, used the same type of dispersion with 34 wt% highly enriched U_3O_8 (10). In addition to the U_3O_8-Al dispersions, UAl_x-Al dispersions were commonly used (x is approximately 3), all these fuels systems still using highly enriched uranium. At this time, in late 70's, the highest uranium density obtained inside the fuel was 1.7 gU/cm^3, which was quite well qualified.

Since highly enriched uranium was easily obtainable in the 70's, the commercial reactors that were using the low-enriched uranium started gradually to convert their cores to highly enriched fuel. Thus, it reached a total of approximately 156 research reactors in 34 countries using highly enriched uranium, resulting in an annual circulation of approximately 5000 kg of this material (11). In 1977, arose again the concern about the proliferation risk associated with loss of fuel during manufacture, transport and storage, leading to restriction by the U.S. government's sale of uranium with high enrichment (above 90 wt% ^{235}U) and producing an impact on the availability and use of the highly enriched fuel for research reactors.

From 1978 programs, it was established for the enrichment reduction, aimed at developing the technology base for replacement of highly enriched uranium by low-enriched uranium (less than 20 wt% ^{235}U) in research reactors. The main program, still active today, is the *RERTR Program* (Reduced Enrichment for Research and Test Reactors), which aims to develop the technology necessary to convert the reactors that use highly enriched uranium (= or > 20% ^{235}U) by low-enriched uranium (less than 20% ^{235}U). During the existence of this program more than 40 research reactors have been converted. At this time, the decrease of enrichment has demanded an effort bigger that previously, because, in most high power research reactors, which are designed to operate in extremes, this substitution involved the development and qualification of new fuels with maximum possible concentration of uranium, which limits are imposed for manufacturability and performance under severe and prolonged irradiation.

In this context, the developments were based initially on increasing the concentration of uranium in the fuel currently used at the beginning of RERTR program, until the practical limit of 2.3 gU/cm^3 in the case of UAl_x-Al and 3.2 gU/cm^3 in the case of U_3O_8-Al. Also, an effort was made in developing new fuels that would allow obtaining uranium densities of 6-7 gU/cm^3, well above the density that can be achieved with the UAl_x-Al and U_3O_8-Al fuel. The development of new fuels would allow the conversion to low enriched from virtually all existing research reactors.

High density of uranium in the dispersion can only be achieved by using the dispersion of fissile compounds with high uranium content. Figure 1 shows the potential of various uranium compounds. The technological limit for the use of dispersions is 45% by volume of fissile material dispersed, since it must be kept a solid aluminum matrix as dispersant. The uranium silicides and U_6Fe compounds were initially considered promising.

The problem encountered in using these intermetallics with high concentrations of uranium as fissile material in the form of dispersions in aluminum is related to its dimensional stability during operation, leading to swelling of the fuel plates and therefore the problems that compromise the thermohydraulic security of the reactor. In mid-1988, based on results from irradiation tests (12) (13), the U_3Si_2-Al based dispersion fuel was qualified by the U.S. Nuclear Regulatory Commission and released for sale with uranium densities up to 4.8 gU/cm^3, with a swelling consistent with the commonly used dispersions (14).

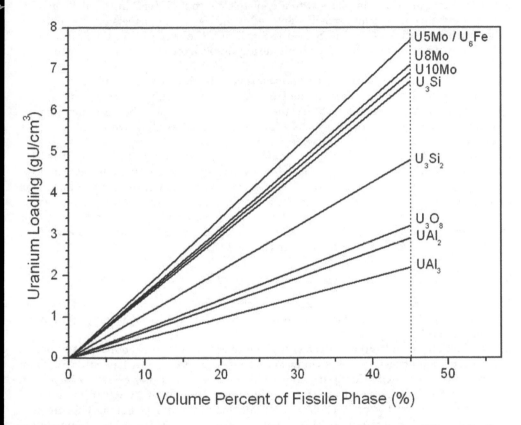

Fig. 1. Density of uranium in terms of concentration of dispersed phase for different fissile uranium compounds.

Research continued aiming at the use of intermetallic with even higher concentrations of uranium, such as U_3Si, U_3SiAl and U_6Fe as fissile material in the form of dispersions in aluminum. However, results of irradiation tests showed an unacceptable dimensional stability of these new fuels. Due to its high concentration of uranium (96 wt%) the U_6Fe was mainly considered (15), and the research was virtually abandoned in 1986 due to high swelling observed in irradiation tests, coupled with the promising results obtained with the U_3Si_2-Al dispersion, being considered a viable alternative (16).

Only through the use of U_3Si_2 as fissile material in the dispersion with aluminum was not possible to convert all research reactors. Many research reactors are awaiting a high-performance technology solution finale, needing a uranium density of 6-9 gU/cm^3. In an effort to convert these reactors, other high density fuels has been studied, including dispersions based on U-Mo, U_3SiCu, $U_3Si_{1.5}$, $U_3Si_{1.6}$, $U_{75}Ga_{15}Ge_{10}$, $U_{75}Ga_{10}Si_{15}$ and uranium nitrides. Still, innovative manufacturing techniques has been investigated, which are based on hot isostatic compaction (HIP - Hot Isostatic Pressing) or increasing the volume fraction of U_3Si_2 beyond 50% (the limit currently accepted for this technology is 45%) or using wires of U_3Si and/or $U_{75}Ga_{10}Si_{15}$ and/or $U_{75}Ga_{15}Ge_{10}$ metallurgically bonded with aluminum in a geometry such that result in plates with a density close to 9 gU/cm^3 in the fuel core. The nearest alternative to be commercially deployed is the UMo alloy dispersion in aluminum, which enables to achieve the density near to 8 gU/cm^3. The performance under irradiation of this type of fuel is being tested with promising results. However, it is not a commercial fuel yet.

Thus, currently the most advanced manufacturing technology commercially available for the MTR type fuel plates is based on the U_3Si_2-Al dispersion, with a concentration of U_3Si_2 resulting in a uranium density into the fuel meat of 4.8 gU/cm^3. The next commercially available technology will probably use a dispersion of UMo alloy with 7-10 wt% Mo, resulting in a uranium density of between 6 and 8 gU/cm^3.

Each type of MTR fuel element is produced in accordance with a manufacturing specification and a set of manufacturing drawings agreed between the fabricator and the reactor operator or his representative. The specification sets down the scope and general conditions, the requirements of manufacturing method, together with the inspection requirements and acceptance criteria. In addition to the specification, an inspection schedule is normally produced which includes all of the supporting documentation such as the inspection and record sheet and certification (17; 18; 19; 20; 21).

3. Production of uranium tetrafluoride

The UF_4 has a specific role in nuclear fuel technology. It is an important intermediate product, being the basic substance to produce either uranium as metal (U^o) or uranium hexafluoride (UF_6) (17).

Uranium tetrafluoride (UF_4) is a green crystalline solid that melts at about 96°C and has an insignificant vapor pressure. It is slightly soluble in water. UF_4 is less stable than uranium oxides and produces hydrofluoric acid in reaction with water; thus it is a less favorable form for long-term disposal. The bulk density of UF_4 varies from about 2.0 g/cm^3 to about 4.5 g/cm^3 depending on the production process and the properties of the starting uranium compounds. Uranium tetrafluoride (UF_4) reacts slowly with moisture at room temperature, forming UO_2 and HF, which are very corrosive.

In principle, several other compounds may also be used for the production of metal and hexafluoride uranium, however, the use of UF_4 is prescribed by technological and economic considerations. It is considerably easier to obtain metallic uranium from UF_4, due to the reactivity of UF_4 mixture with reducing agent (mainly Ca and Mg) with large thermal outcome, which makes easy the production of uranium ingot.

According to the production process, the UF_4 must have certain specifications in regard to its purity. The content of uranium oxides and uranyl fluoride (UO_2F_2) may vary and also its density and its granulometric composition. The major technical requirement for tetrafluoride is observed during metallic uranium fabrication. It must contain at least 96% of tetrafluoride, virtually free of impurities. It should be anhydrous and having sufficiently high density.

When the reduction process to produce metallic uranium is performed at higher pressures and lower temperatures, normal tolerances up to 4% of $UO_2 + UO_2F_2$ should be reduced. It is recommended that the tapped density of loose UF_4 should be greater than 1 g.cm^{-3}.The good quality of the metallic uranium to be produced should have the UF_4specification as displayed in Table 1.

Elements	Al	B	Cd	C	Co	Cr	Cu	Fe	Mn	Ni	Si
% in mass	70	0,2	0,1	150	5	25	40	75	15	40	30

Table 1. Specification of the limits for main impurities in UF_4

In the case the oxide content is high, there would be larger losses of metal with the slag. As the reaction develops high amount of heat, it should be avoided evolution of tetrafluoride volatile components such as water and ammonia. During smelting, the metal is slightly contaminated with impurities from reducing agent and crucible. For this reason, UF_4 should be pure enough to allow slight contamination degree during this process. It must also be sufficiently dense. The load consists of blending of UF_4 powder and chips of calcium or magnesium tetrafluoride. The higher the density of UF_4, the greater the density of the load, and the greater the amount of heat involved per unit volume of the furnace (17).

The production of uranium tetrafluoride can be made by several processes which are divided into two groups, namely dry (fluorination of uranium oxide or hexafluoride reduction) and aqueous (preparation of UF_4 from U^{+6} salt) pathway (18) (19) (20) (21) (22).

The first task of obtaining UF_4 were carried through water (22) (23) by the end of the 19th century, and from an industrial standpoint that prevailed till the beginning of 20th century. The process essentially comprises the steps of reducing the uranium contained in uranyl fluoride solutions, uranyl chloride or uranyl sulfate up to its tetravalent state, followed by UF_4 precipitation by adding hydrofluoric acid.

With the development of dry processes, the aqueous processes were abandoned because they had difficulties in filtration, washing and drying, in spite of their simplicity and safety. Nowadays, the production via aqueous route is only used in plants to produce UF_4 for small quantities, which is the present experience of IPEN in producing LEU UF_4. Nevertheless, IPEN also developed the Brazilian technology for dry route.

3.1 Procedures for obtaining UF_4 via wet process
3.1.1 Preparation of UF_4 from salts U^{+6}

The UF_4 preparation methods through water have been developed mostly by the British and its modifications were based on work done by Bolton in 1866 (24; 25)

Essentially, the process consists in reducing the uranium, contained in solutions of uranyl fluoride, uranyl chloride and uranyl sulfate to the tetravalent state and the precipitation of uranium tetrafluoride by adding hydrofluoric acid. Several compounds of uranium have been used as starting materials and various reducing agents have been used. An overview of the process can be obtained from the reaction of uranyl fluoride with stannous chloride and sodium hyposulphide.

$$UO_2F_2 + SnCl_2 + 4HF \leftrightarrow UF_4 + 2H_2O + SnCl_2F_2 \qquad (1)$$

$$UO_2F_2 + Na_2S_2O_4 + UF_4 + 2HF \leftrightarrow Na_2SO_3 + SO_2 + H_2O \qquad (2)$$

An alternative to this process is the replacement of the electrolytic reduction by reducing agents that prevents the possible contamination with the reducing agent. This process has been adopted in countries like USA, Spain, Australia, Japan, Canada, England, South Africa and India (26; 27; 28; 29; 30; 31; 32). For the production of UF_4 with nuclear purity from UO_2F_2 acids solutions, some fundamental stages are required such as obtaining the solution, reduction to uranium valence and precipitation of the formed U^{+4}. These stages are shown in Figure 2, as schematized operations.

3.1.2 Obtaining UO_2F_2 solutions

Uranium hexafluoride is a crystalline substance at normal pressure and temperature conditions. At the temperature of 900C under a pressure of 3kgf/cm2, UF_6 becomes gas and when it is injected into water, it hydrolyzes immediately according to the following:

$$UF_6 + H_2O \rightarrow UO_2F_2 + 4HF \tag{3}$$

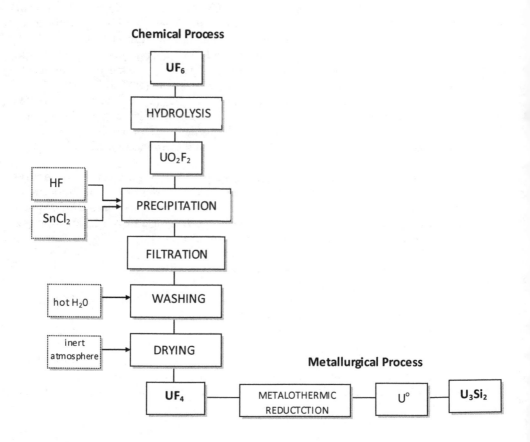

Fig. 2. Wet Process to produce UF_4

Table 2 shows the chemical characteristics of UO_2F_2 solution obtained from UF_6 hydrolysis.

Uranium (g/L) 60									
Fluoride (g/L) 17									
Metallic impurities (g/mL)									
Cd	B	P	Fe	Cr	Ni	Mo	Zn	Si	Al
<0.1	0.2	<100	1500	100	40	<2	100	300	40
Mn	Mg	Pb	Sn	Bi	V	Cu	Ba	Co	
10	15	<2	<2	<2	<3	3	1	<10	

Table 2. Chemical characteristics of UO_2F_2 solution

3.1.3 Chemical reduction of UF_6 to UF_4

Uranium in its tetravalent state is very important in different technological processes. Essentially, the preparation process (aqueous way) from solutions containing uranyl ion (hexavalent) involves the reduction towards tetravalent state, and later precipitation as UF_4 using HF solution. In aqueous solutions, these reductions can be carried out by chemical, electrochemical or photochemical methods.

All the trials for the preparation of UF_4 using chemical reduction have been carried out using UO_2F_2 solution inside a stainless steel reactor, coated with Teflon. The solution has been heated under continuous stirring to reach a temperature set, and the reducing agent has been added. Next, the precipitating agent solution is slowly added to UO_2F_2 in solution with hydrofluoric acid (HF). Tests have been carried out using some reducing agents, such as $SnCl_2$, $CuCl$, $FeCl_2$, $Na_2S_2O_4$.

$$UO_2F_2 + SnCl_2 + 4HF \rightarrow UF_4 + SnClF_2 + 2H_2O \tag{4}$$

$$UO_2F_2 + 4HF + Fe \rightarrow UF_4 + FeF_2 + 2H_2O \tag{5}$$

$$UO_2F_2 + CuCl + 4HF \rightarrow UF_4 + CuClF_2 + 2H_2O \tag{6}$$

$$UO_2F_2 + Na_2S_2O_4 + 2HF \rightarrow UF_4 + Na_2SO_3 + H_2O \tag{7}$$

Upon UF_4 precipitation the suspension is left in rest up to reaching room temperature. After over 12 hours, it was performed the solid/liquid separation by vacuum filtration, washing and drying in a muffle kiln. The salts obtained were all identified as being uranium tetrafluoride. According to the results shown in Figure 3, it is evident that, from all used reducing agents, only $SnCl_2$ and $FeCl_2$ have shown significant results in regards of getting UF_4. Nevertheless, $SnCl_2$ is more consistent reducing agent at higher temperature of process.

The influence of the temperature upon UO_2F_2 and UO_2 contents in obtained UF_4 is shown in Figure 4. It was employed $SnCl_2$ as the reducing agent in this study to precipitate UO_2F_2 solution. The residual moisture is dried at 130°C. The tin content in all obtained UF_4 has shown to be in the range of 0.15 – 0.15%.

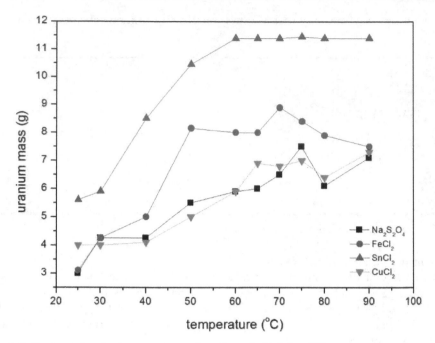

Fig. 3. Influence of reducing agent as a function of obtaining UF₄

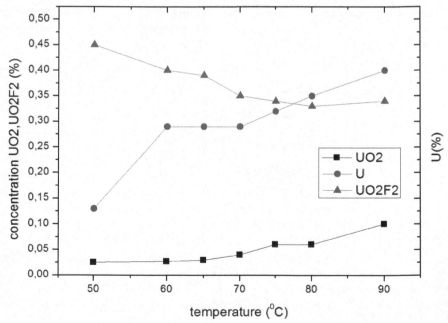

Fig. 4. Influence of the temperature as a function of the contents of $UO_2 F_2$ and UO_2 in UF₄

3.1.4 Obtaining UF4

As shown previously, the process for obtaining UF_4 by reduction precipitation using $SnCl_2$ had the best results and achieved an yield of 98% of UF_4 precipitation. The precipitation with HF solution is relatively slow and tends to accelerate as the temperature rises (17; 18). This is important, since it avoids excessive precipitate hydration and facilitates the sedimentation, filtration and drying operations. The full reaction is represented by:

$$UO_2F_2 + SnCl_2 + 4HF \rightarrow UF_{4\,pp} + SnCl_2F_2 + 2H_2O \qquad (8)$$

During the uranium processing stages, the goal is to achieve an end product with high purity and showing physical and chemical characteristics appropriate for the preparation of nuclear fuel.

Table 3 lists the suitable chemical and physical characteristics of UF_4 for a later reduction to obtain metallic uranium.

	at 130°C			inert atmosphere at 400°C				
Uranium (%)	74.20			75.0				
Fluoride (%)	24.60			27.90				
UF_4 (%)	97.50			99.85				
UO_2F_2 (%)	0.29			0.34				
UO_2 (%)	0.06			0.29				
HF(%)	0.23			0.12				
Moisture (%)	0.33			<0.03				
Crystallization H_2O	4.50			<100				
Met. Impurities	Fe	Cr	Ni	Mo	Al	Mn	Cu	Sn
$(\mu g/g)$	<20	<10	<10	<5	<10	<5	<5	0,1
Density (g/cm^3)	6.70							
Granulometry (m)	15.0							
Specific Surface (m^2/g)	0.21							

Table 3. Chemical and Physical Properties of UF_4 produced by an aqueous route

3.1.5 Preparation of UF4 from UO2

The UF_4 obtained by reaction with UO_2 with hydrofluoric acid is easily made. The reaction can be summarized as follows:

$$UO_2 \,(s) + 4HF \,(aq) \leftrightarrow UF_4 \,(s) + 2H_2O \qquad (9)$$

This process has some advantages over the other processes. Since the reaction occurs at low temperatures, the reactor can be constructed using materials as polyethylene, polypropylene or carbon steel with plastic coating, while other processes require equipment built with metal (monel, inconel, nickel) which increases the cost of a plant.

In Figure 5, the x-ray diffractogram spectra are presented for UF_4 produced by the method via NH_4HF_2 (bifluoride route) and by aqueous route. Typical SEM image of precipitated UF_4 is presented in Figure 6. It displays a granular structure with relevant amount of porosity.

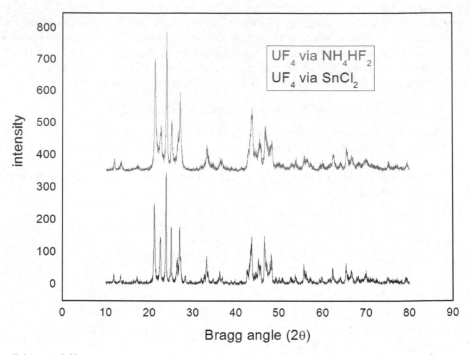

Fig. 5. X-ray diffraction pattern of UF₄ produced by the bifluoride route and from the aqueous route.

The UF₄ fabrication using fluorination media with ammonium bifluoride is perfectly feasible. The ammonium bifluoride is a by-product effluent generated during the UF₆ conversion to AUC[1]. UF₄ obtained by this route has the same crystalline structure presented by the aqueous process, as demonstrated by the x-ray spectrum. Besides, it has the correct chemical and physical characteristics for metallothermic production of metallic uranium. Even presenting a lower relative tapped density; this property will not be a problem, because this is an alternative process that has as main goals the recovery of uranium, ammonium and the fluorides of the liquid effluents generated in the process of UF₆ reconversion. This UF₄ will be lately diluted in the UF₄ charges produced by the aqueous route. The development of this process (bifluoride route) not only provides an efficient process for uranium recovery from secondary sources, as also eliminates the environmental pollution by discarding the bifluoride. It also provides a chemical compound with chemical and physical characteristics very similar to the aqueous route (SnCl₂).

---

[1] Ammonium uranyl carbonate ($UO_2CO_3 \cdot 2(NH_4)_2CO_3$) is known in the uranium processing industry as AUC and is also called uranyl ammonium carbonate. Ammonium uranyl carbonate is one of the many forms called yellowcake in this case it is the product obtained by the heap leach process. This compound is important as a component in the conversion process of uranium hexafluoride (UF_6) to uranium dioxide (UO_2). In aqueous process uranyl nitrate is treated with ammonium bicarbonate to form ammonium uranyl carbonate as a solid precipitate and ammonium bifluoride as by-product (41).

Fig. 6. SEM image of some UF_4 particles, produced by the bifluoride(a) route e via SnCl2 (b).

3.2 Procedures for obtaining UF_4 by dry process
3.2.1 Preparation of UF_4 by fluorination of UO_2

The achievement of UF_4 by this process was adopted in Canada, France, the former Czechoslovakia, South Africa, United States, Portugal, Brazil, Germany and Sweden (17; 21; 23).

The sequence of operations is to reduce UO_3 by hydrogen, followed by treatment with HF resulting UO_2 anhydrous at atmospheric pressure.

$$UO_3 \text{ (s)} + H_2 \text{ (g)} \leftrightarrow UO_2 \text{ (s)} + H_2O \text{ (v)} \qquad (10)$$

$$UO_2 \text{ (s)} + 4HF \text{ (g)} \leftrightarrow F_4 \text{ (s)} + 2H_2O \text{ (v)} \qquad (11)$$

The reduction of UO_2 is performed at temperatures of 500-700°C. Another alternative is the reduction of U_3O_8 recommended when you have storage problems UO_3, being extremely hygroscopic.

$$U_3O_8 \text{ (s)} + 2H_2 \text{ (g)} \leftrightarrow 3UO_2 \text{ (s)} + 2H_2O \text{ (v)} \tag{12}$$

In such a process is commonly used the moving bed or fluidized bed reactor type.
Preparation of UF_4 by reaction of the UO_3 with NH_3 and HF gaseous
The process consists of only one step to produce UF_4. The mixture consisting of NH_3 and HF is treated with UO_3 at 500-700°C. This reaction is fast and produces high purity UF_4:

$$3UO_3 + 2NH_3 + 12 \text{ HF} \leftrightarrow 9H_2O + N_2 + 3UF_4 \tag{13}$$

The UF_4 fabrication by the reaction of uranium oxides with fluorinated hydrocarbons (freon) is as follows:

$$2CF_2 Cl_2 + UO_3 \leftrightarrow UF_4 + CO_2 + Cl_2 + COCl_2 \tag{14}$$

The literature shows results of reactions of different freons with uranium oxides UO_2, U_3O_8 and UO_3 (27; 29; 33). The reactors used in this process cannot be constructed using nickel, copper, platinum and stainless steel, since they undergo chemical attack of reagents, besides this reaction promotes pyrolysis under carbon presence. The reactors are constructed with graphite or calcium fluoride, which may cause contamination to the obtained UF_4. The advantages of this method are equipment simplicity and the possibility of applying this reaction to all the uranium oxides.

3.2.2 Preparation of UF₄ from metallic uranium or uranium hydride (UH₃)

By fluoridation at high temperatures uranium metal can be quickly converted into uranium tetrafluoride by the reaction below:

$$U + 3/2H_2 \overset{250°C}{\leftrightarrow} UH_3 \tag{15}$$

$$UH_3 + 4HF \overset{200°C}{\leftrightarrow} UF_4 + 7/2H_2 \tag{16}$$

Uranium metal is industrially manufactured from UF_4. In the absence of advantage in obtaining first elemental uranium and transform it into UH_3, then get to UF_4.

3.2.3 Procedures for obtaining UF₄ by dry ammonium bifluoride with (NH₄HF₂)

The fluorination of UO_2 is made with NH_4HF_2, a white solid; it has low vapor pressure and can be operated freely since it is non-toxic. Initially, UO_2 is mixed with bifluoride, 20% above the stoichiometric amount. The bifluoride crystal is easily crushed and the mixture of $UO_2 + NH_4 HF_2$ is made in a monel 400 container to prevent contamination.

The conversion of bifluoride at room temperature occurs after approximately 24 hours, although under such conditions the water formed in the reduction may be retained in the precipitate. The elimination of NH_3 and water is facilitated by the reaction of UO_2 and NH_4HF_2 at 150°C:

$$2UO_2 + 5NH_4 HF_2 \leftrightarrow 3NH_3 + 4H_2O + 2NH_4 UF_5 \tag{17}$$

At this temperature, only 8 hours are necessary to promote the fluorination. The material is loaded into an aluminum container with calcium fluoride and heated inside a furnace. The furnace is fitted with a condensing tube with a relief valve, which releases the water and ammonia from the fluoridation reaction to a reservoir and retains the excess of sublimed bifluoride.

During the fluorination and/or decomposition, the formation of UO_2F_2 probably occurs. This is a significant happening, since it may reduce the efficiency of reduction in the next step.

In a second step of the process, under vacuum distillation, NH_4UF_5 is decomposed in UF_4 with the NH_4F by this reaction:

$$NH_4 UF_5 \leftrightarrow UF_4 + NH_4F \tag{18}$$

3.2.4. Preparation of UF_4 by the reaction of ammonium bifluoride with UO_3

The UF_4 can be prepared by reaction of ammonium fluoride or bifluoride with UO_3 according to the equation:

$$3UO_3\ 6NH_4HF_2 +9H_2O \leftrightarrow 3UF_4 +4NH_3 +N_2 \tag{19}$$

Although the United States have been among the first to study the process (34) Canada is the country that developed this process (35; 36)

4. Production of metallic uranium

There are several possibilities to produce metallic uranium (41; 26; 42). Magnesiothermic reduction of UF_4 is one of them and it is a known process since early 1940's (7; 8). The IPEN technology uses this route in 1970-80's for production 100kg ingots of natural uranium. For LEU U-production, it is necessary to handle safe mass (less than 2.2 kg U), to avoid possible criticality hazards. IPEN presently produces around of 1000g LEU ingots via magnesiothermic process and in future may produce 2000g or more. This range of uranium weight is rather small if compared to big productions of natural uranium. Metallic uranium is reported (9) to be produced with 94% metallic yield when producing bigger quantities. The magnesiothermic process downscaling to produce LEU has small possibilities to achieve this higher metallic yield. This is due to the design of crucibles, with relatively high proportion of surrounding area, which is more prone to withdraw evolved heat from the exothermic reaction during uranium reduction. Normally, calciothermic reduction of UF_4 is preferred worldwide, since the exothermic heat is higher (-109.7 kcal/mol) compared to smaller amount of -49.85 kcal/mol using magnesium as the reducer (10). Nevertheless, IPEN chose magnesiothermic because it is easier to be done, avoiding no handling of toxic and pyrophoric calcium. Moreover, the magnesiothermic process is cheaper, so, it brings economical compensation for its worse metallic yield than calcium reduction process. In addition, the recycling of slag and operational rejects is highly efficient and there are virtually insignificant LEU uranium is lost (23).

The magnesiothermic reaction is given by:

$$UF_4 + 2Mg = U + 2MgF_2\ \Delta H= -49.85\ kcal/mol\ (at\ 640°C) \tag{20}$$

As magnesium thermodynamics is less prompt to ignite than calcium, the batch reactor is heated up to the temperature around 640°C. The routine shows that this ignition normally happens some degrees bellow this temperature (9). Nevertheless, several reactions may occur during heating of the UF_4+Mg load. Moisture is normally present in the charge, either caught during UF_4 handling after drying or during crucible charging. During heating, as the temperature crosses the water boiling point (>100°C), all moisture becomes water vapor. This vapor not only bores its passage through the load but easily oxidize the reactants in this pathway by the following reactions (30):

$$UF_4 + 2H_2O \rightarrow UO_2 + 4HF \tag{21}$$

$$2UF_4 + 2H_2O \rightarrow 2UO_2F_2 + 4HF \ \text{(via } UF_3(OH) \text{ and } UOF_2 \text{ steps)} \tag{22}$$

As the loading of the charge is not fully sealed to avoid atmosphere contact, some O_2 is entrapped in the system, leading also to reactants oxidation by:

$$2UF_4 + O_2 \rightarrow UF_6 + UO_2F_2 \tag{23}$$

Producing some UF_6 that transforms into UO_2F_2 by the following reaction:

$$UF_6 + 2H_2O \rightarrow UO_2F_2 + 4HF \tag{24}$$

and also occurring magnesium oxidation (very fast above 620°C) by:

$$2Mg + O_2 \rightarrow 2MgO \tag{25}$$

The presence of the UO_2 and UO_2F_2 in the produced UF_4 accumulates with previous oxidized ones during the dehydration. All these compounds formation worsens the metallic yield of uranium production.

In this work, it is discussed the effect of LEU UF_4 precipitated via hydrolyzed UF_6 and its potential variability in reactivity. The chemical UO_2F_2 residual content in dried UF_4 is also analyzed for its potential relevance in the uranium production. The tapped density of dehydrated and loaded UF_4 is also commented as affecting the reactivity process of uranium production. The magnesiothermic ignition is also analyzed since the heating time of the charge may affect the reactivity of the load. The reaction sequence after ignition is theoretically proposed as a possible sequence of chemical and physical events. The evidences in the slag solidification on crucible wall, during the reaction process to reduce UF_4 towards U°, is very enlightening to guide towards the interpretation of the reaction blast.

The IPEN's magnesiothermic reduction process of UF_4 to metallic uranium (in the range of 1000g) could be synthesized as:

1. In preparation for the mass reduction of a single batch, it is used with a standard charge of reactants of 1815 ± 5g of the mixture Mg + UF_4 (1540 ± 1g LEU UF_4) containing 15% excess of stoichiometric Mg content. For purpose of homogenization, the charge of UF_4 + Mg is divided into 10 layers, which are tapped one by one inside the crucible. All this operation is carried out inside a glovebox to prevent nuclear contamination. This sequence is illustrated in Figure 7.

2. After placing the reactants inside the graphite crucible, a variable amount of CaF_2 is tapped over the UF_4+Mg load in the crucible to fully complete the reaction volume. This amount is dependent on tapped density and UF_4+Mg blending, which varies in function to UF_4 fabrication. The crucible is made of fully machined graphite volume with enough resistance to produce safe nuclear uranium amount around 1000g. This crucible was designed to withstand the blast impact of metallothermic reaction, as well as thermal cycles of heating and cooling without excessive wear in order to be used in several batches.

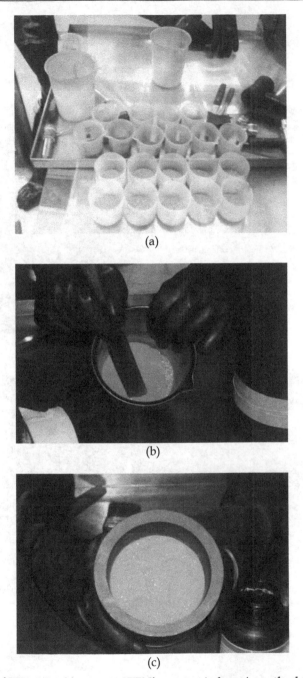

Fig. 7. Sequence of UF4+Mg charging in IPEN's magnesiothermic method to produce metallic uranium. (a) 10 layer preparation of UF_4 (green) and Mg (metallic bright); (b) blending of material; (c) full charge after tapping the 10 layers.

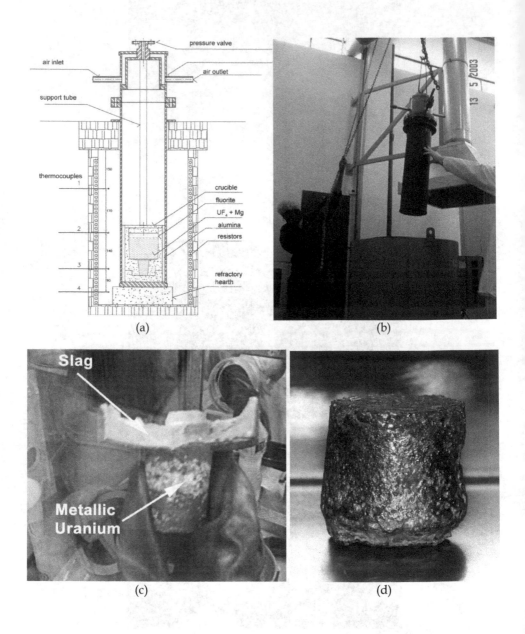

Fig. 8. (a) Schematic drawing of pit furnace, reactor vessel and crucible; (b) Charging of the reactor vessel inside the pit furnace; (c) Raw metallic uranium and upper deposited slag after removing from the crucible; (d) Metallic uranium after cleaning.

3. After closed with the top cover, the crucible is inserted inside a stainless steel cylindrical reactor vessel, made of ANSI 310, which allows argon fluxing during batch processing (1 L/min with 2 kgf/cm^2 of pressure). As shown in Fig. 8 (a-b), the whole crucible + reactor are placed in resistor pit furnace with four programmable zones having the possibility of raising the temperature up to 1200°C.
4. The reaction vessel is set to heat up to 620°C. At this level, the reaction ignition is expected. The total heating time and waiting for ignition is about 180 minutes from heat time to temperature setting point.
5. The reaction of UF_4 with Mg produces an intense exothermic heat release inside the crucible. It is considered as an adiabatic reaction. It produces metallic uranium and MgF_2 slag in liquid form. Both products deposit in the crucible bottom are easily taken apart after opening the crucible. Some products project over the crucible wall and freeze there.
6. This full reaction happens in a noticeable time between 800 and 1200ms from ignition to final deposit. This control is measured by sound waves, using an accelerometer.
7. After the reaction, 10 minutes is awaited for full solidification of reaction products inside the furnace. Then the furnace is turned off and the reactor vessel is lifted out of the furnace. There is a 16 hours for cooling before its opening. This avoids firing of metallic uranium in contact with atmosphere.
8. The disassembling of reduction set is performed inside a glove box. The top and bottom covers of the crucible are removed. By means of rubber soft hammering, it is able to withdraw the uranium ingot. The MgF_2 slag is removed by mechanical cleaning. The metallic uranium is pickled in nitric acid 65%vol and the final mass of metallic uranium is measured and its density evaluated by Archimedes' method.

5. Production of uranium silicide

The intermetallic U_3Si_2 is produced from metallic uranium (47). This alloy is produced from a uranium ingot and hyperstoichiometric silicon addition (7.9% Si). The induction furnace (15 kW) should be submitted to 2.10^{-3} mbar vacuum and flushed with argon-atmosphere. Then the melting is carried out. The blend is molten inside an induction furnace using zirconia crucible reaching more than 1750°C, as this intermetallic requests this level of temperature to be properly homogenized before solidification. No other crucibles, than a zirconia one could bear the aggressive environment created by uranium attack on linings. The load arrangement of uranium and silicon, as shown in Fig. 9, is then charged inside the crucible. It was planned to help the sequence of melting during the several stages that passes the alloy formation until reaching the final intermetallic composition. The quality of this intermetallic produced in this way normally meets the requirements as nuclear material. The X-ray diffractogram (Fig. 9) confirms the necessary proportion of phases presents in the produced powder of this alloy, which should be more than 80wt% of crystalline phases. As rule of thumb, the chemical amounts of boron, cadmium, cobalt, lithium should be less than 10μg/g individually. The other may reach hundreds of μg/g up 1000 μg/g. Carbon could reach up to 2000 μg/g. Isotopic concentration of [235]U is 19.75±0.20wt%. The required density is 11.7g/cm^3.

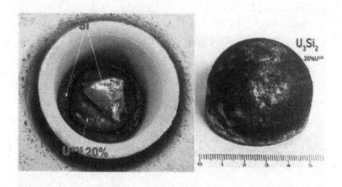

Assembling of U₃Si₂ crucible load

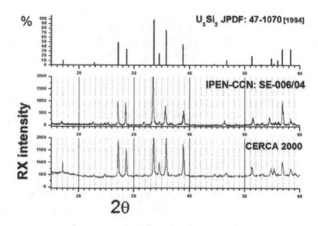

X-ray diffractogram of U₃Si₂ produced in Ipen compared

Fig. 9. Crucible arrangement of before melting to produce the intermetallic. U₃Si₂ product and its x-ray diffractogram results compared to CERCA product and JPDF 47-1070 for pure U₃Si₂.

6. Production of MTR nuclear fuel

The reference industrial process to produce plate-type fuel involves roll-milling together the fissile core, or fuel meat (a blend of an uranium compound and aluminum powders), and the cladding (aluminum alloy plates). This process can draw on considerable feedback from experience, since nearly all research reactors use this type of fuel. The process has seen large-scale implementation with NUKEM, in Germany, UKAEA, in the United Kingdom, CERCA, in France, and Babcock, in the United States.

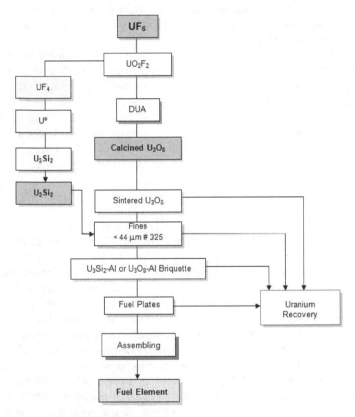

Fig. 10. Fabrication process of silicide fuel elements.

In general, the MTR type fuel element fabrication process using silicide (U_3Si_2) can be divided into the following main steps: hydrolysis of UF_6 through its reaction with water; production of uranium tetrafluoride (UF_4); production of metallic uranium; U_3Si_2 powder production from uranium metal; production of fuel cores from U_3Si_2 and aluminum powders; production of fuel plates with U_3Si_2-Al dispersion; assembling of fuel elements; recovery of uranium; effluent treatment; quality control.

The simplified block diagram of the fabrication process for silicide fuel elements is shown in Figure 10. The manufacturing process of the fuel begins with the UF_6 processing. The UF_6 is enriched to 19,75 wt% ^{235}U, a enrichment level that categorize the fuel as LEU (low enriched uranium). Bellow the main stages of manufacture of such fuel are discussed.

6.1 Fuel cores production from U_3Si_2 and aluminum powders

The U_3Si_2 ingot produced in the previous step is transferred to a glove box with inert atmosphere of argon, since the U_3Si_2 is pyrophoric. Inside the glove box, the ingot is subjected to a preliminary grinding, resulting in granules less than 4 mm in size with the smallest fraction of fines (< 44 μm) possible. This operation is performed with the aid of a manual crusher. After doing the preliminary grinding, the material is placed directly on a set of sieves, and then sieved by hand. The sieve set comprises a coarse sieve with 4 mm opening, a fine sieve with 150 μm opening and a background compartment. The granules with a diameter greater than 4 mm are crushed again. The granules with size between 4 mm and 150 μm are collected for final grinding and particles smaller than 150 μm are collected separately for particle size classification.

The U_3Si_2 obtained after the preliminary grinding is manually milled again. The material collected during the preliminary grinding (between 4 mm and 150 μm) is processed in this step. The grind is done carefully, with intermediate sieving, to classify the powder in the range from 150 to 44 μm. The specification allows 20 wt% fines fraction (below 44 μm) as maximum. The fraction above the specification (150 μm) is sent back to the final grinding system. The fraction inside the specified range (between 150 μm and 44 μm) is collected and stored. The fraction of fines (< 44 μm) is collected and stored separately. The glove box contains a vibrating screening machine, which performs the separation of three size fractions of silicide powder, above 150 μm, between 150 and 44 μm and bellow 44 μm. The batch U_3Si_2 powder composition is adjusted to have maximum fines content in the level of 20 wt%, as specified.

The next process step is the fabrication of the fuel cores, which will form the core of the fuel plates, or fuel meats. The core of the fuel plate contains U_3Si_2 as the fissile material. This core is fabricated by means of powder metallurgy techniques and is normally called briquette or fuel compact. Initially, the mass and composition of the briquette are calculated based on the analyzed values of total uranium and isotope enrichment of the U_3Si_2 powder. The criterion for calculating the briquette mass is the amount of the isotope [235]U specified for the fuel and the dimensions of the briquette. Based on the calculated mass of the briquette, the silicide Al powders mass are determined separately and mixed together to ensure that the specified [235]U amount is uniformly distributed. These charges are cold pressed to form the fuel compacts, and the briquettes are measured and weighed. The final dimensions of the fuel meat in the finished fuel plate are set by specification and the volume of the briquette is calculated from these data by their values of thickness, width and length. The thickness of the briquette is obtained by multiplying the specified thickness of the fuel meat by the deformation dimension resulted after rolling operation, assuming zero enlargement. The core content of voids depends only on the volume fraction of fuel powder content. To optimize the final geometry of the rolled core, the briquette has rounded corners, and the volume of the corners is included in the calculation of volume.

The difference between the volume of the briquette, obtained as described above, and the volume of the fuel powder, as determined by the division between the mass of the powder and its density, determines the amount of aluminum powder to be added to the mass of the briquette. As the theoretical density of the system cannot be achieved during the compaction of the briquette, the volume of aluminum is reduced by the amount of pores that remain after pressing. The total mass of the briquette is given by the calculated mass of fissile material powder added to the calculated mass of aluminum powder.

According to the calculation for the masses of U_3Si_2 and aluminum powders, the charges for pressing are weighed separately. The weighing is carried out in glass bottles specially designed for installation in a homogenizer. Once the powders are weighed, the charge is mixed inside a glove box with inert atmosphere. This blending ensures that the specified amount of ^{235}U is homogeneously distributed throughout the briquette to be pressed. The weighing operation is performed carefully and, after homogenization, the cautious handling of the charge is critical to avoid segregation.

The homogenization operation is performed using a special homogenizer with a capacity for simultaneous mixing of eight charges. The duration of homogenization is 120 minutes under rotation of 36 rpm and angle of 45°. To prepare the briquettes, the homogenized charges are pressed at room temperature using a hydraulic press with capacity for 700 tons, which is placed in a glove box. The pressing pressure is adjusted to get the desired thickness, keeping the residual porosity from 5 to 7% by volume. The bottle containing the homogenized charge is transferred from the glove box used for homogenization to the glove box used for pressing. Within this glove box, the charge of a briquette is emptied into the die cavity with the inferior puncture initially raised. The powder is placed in layers with the aid of a special smoother to prevent segregation and to minimize the variation of the thickness of the briquette, lowering the punch inferior gradually until all the charge is loaded, when the punch is fully lowered to its position during the pressing. Then, the superior punch is inserted and pressure is applied and maintained for 15 seconds. The entire array is then opened to eject the briquette and the punch superior, which is manually removed. The thickness of the briquette is defined based on final specifications valid for the fuel meat. This thickness is theoretically calculated and then adjusted through manufacturing tests.

Immediately prior to the transfer of the briquettes to be used in the manufacture of fuel plates, they are vacuum degassed at 2×10^{-3} torr in a retort. The temperature is 250 °C kept for 1 hour. After remaining inside the degassing retort for the time and temperature specified, the briquette is removed for cooling, keeping the vacuum system working until the room temperature is reached. Thus, the briquettes that will compose the cores of the fuel plates are used in the new phase of processing, or assembling the sets for rolling. Figure 11 illustrates the process for preparing the briquettes and the set.

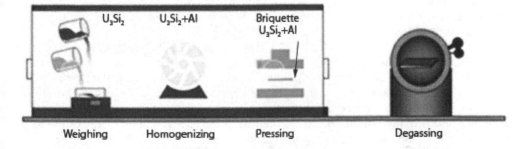

Fig. 11. Process for briquettes preparation and degassing.

6.2 Production of fuel plates with U_3Si_2 – Al dispersion

The technology of fuel plates manufacture adopts assembling and rolling of a set composed by the fuel meat (briquette), a frame plate and two cladding plates. In this way, after the rolling operation, it is fabricated a fuel plate containing inside the fuel meat totally isolated from the environment, which is done through the perfect metallurgical bonding between the core and frame with the claddings. The frame and cladding plates are made from commercial aluminum Al 6061 alloy (48).

In order to prepare the rolling assemblies, the frame plate is heated in a furnace at 440 °C. The cold briquette is then assembled inside the frame plate. Once cooled the frame, the briquette should be perfectly housed and fixed in the frame cavity by mechanical interference. The other cladding plates are placed above and below the frame plate with the core, completing then the assembling to be rolled. This assembly set is then fixed in a rotating welding bench and welded at its edges. The welding is TIG type protected with argon. A continuous welding bead is done on the four corners of the assembly, leaving the ends free in order to allow air to be exhausted in the first rolling pass. Figure 12 illustrates the procedure of preparing the assemblies for rolling.

Fig. 12. Diagram illustrating the assembling of the set core-frame-claddings.

The welded assemblies are properly identified and inserted in a furnace for 60 minutes at a temperature of 440 °C. The hot rolling is performed in several passes following a well-established rolling schedule. The rolling schedule defines thickness reduction per pass in order to control the end defects and the final dimensions of the fuel meat. The rolling schedule is determined by theoretical calculations and empirical data from manufacturing tests and must guarantee the metallurgical bonding and the control and reproducibility of the fuel meat deformation. The rolling mill usually has an accuracy of 0.025 mm and is equipped with rolling cylinders coated with a chrome layer. It is important the perfect lubrication of the rolling cylinders. Between each pass, the assemblies are reheated for 15 minutes. After the final hot-rolling pass, the fuel plates are identified again in the same position of the initial identification in a region outside the fuel meat, using mechanical marker.

After hot rolling, a blister test is performed to test the metallurgical quality of bonding between meat-frame-claddings. The hot rolled plates are heated at 440 °C for 1 hour. After removal from the furnace, the fuel plates are visually inspected for observation and recording of bubbles (47; 48). Fuel plates that present bubbles are registered as reject and forwarded for chemically recover of uranium.

The cold rolling operation is performed in the same rolling mill used in the hot rolling. In this operation the specified thickness is achieved with precision. The total cold reduction is approximately 10% in thickness and is applied in one or two passes. During cold rolling, the length of the fuel meat is checked, ensuring the fulfillment of the specification for the minimum core length and for the thickness of the fuel plate.

After cold rolling, the fuel plates are pre-cut for facilitate handling during the subsequent fabrication operations, as flattening, radiography and final cut. The fuel plates obtained in cold rolling have their surfaces still undulating, requiring a flattening operation. This operation is performed using a roll-flattener, which is basically consisted with a group of flattener cylinders controlled by a position adjustment system to keep the cylinders in a flat position. Only one pass is enough to flatten the fuel plates.

The next step is the final cut of the fuel plate to reach the specified dimensions. This cut is made using a guillotine cutter machine and is oriented by x-ray radiography. This radiography is obtained by using an industrial system set, where the fuel meat can be perfectly positioned inside the fuel plate and, then, the plate receives line tracing to guide the final cut. Next, the fuel plates are degreased in acetone and pickled in a solution of NaOH 10wt% for 1 minute at 60 °C. Then, they are washed in water for 1 minute, neutralized in cold 40wt% HNO_3 for 1 minute, rinsed again in running demineralized water for 5 minutes (spray), washed by immersion in hot demineralized water and dried manually with the aid of hot air blast. Figure 13 shows a drawing of the fuel plate, illustrating its fuel meat. Figure 14 shows the sequence of operations performed to manufacture the fuel plates.

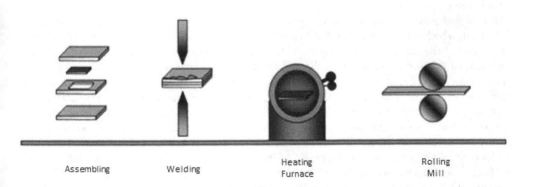

Assembling   Welding   Heating   Rolling
                                     Furnace   Mill

Fig. 13. Illustration of the process for preparing the assemblies and rolling

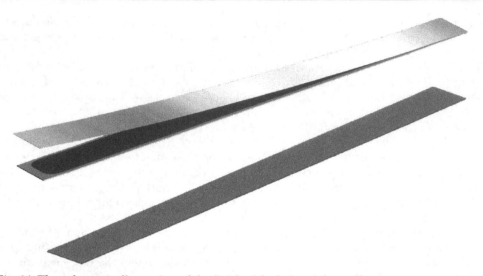

Fig. 14. The schematic illustration of the finished fuel plate (after rolling).

The finished fuel plates are characterized dimensionally, measuring in its length, width and thickness. Fuel plates that do not meet the dimensional specifications are rejected and sent for uranium recovery.

After the final cut, two new radiographs are obtained. The first one aims at checking the position of the fuel meat inside the fuel plate, as well as to verify its dimension, length and width. The second radiography aims to check the uranium distribution homogeneity in the fuel meat and also its integrity, as well as the possible presence of "white spots" and fissile particles outside the fuel meat zone.

To check the reproducibility and stability of the manufacturing process of fuel plates, the residual porosity of the fuel meat of all fuel plates produced are determined using the Archimedes principle.

Every 24 fuel plates produced, one fuel plate is separated to characterize the end defects in the fuel meat, which are basically the cladding thickness reduction in the area of the "dog-boning", inspection of the "diffuse zone" (end of the fuel meat) for studying the "fish tail" defect and to do the final geometry inspection of the fuel meat. This analysis is performed destructively to allow metallographic image analysis. In the case of fuel plate production routine, the quality analysis samples is randomly made (1:20) over all produced plates to check possible defects that do not meet specifications. In case, the sample is rejected then a second fuel plate is randomly taken from the batch and is destroyed to be examined. If this second sampled plate also proves defective then the entire batch is rejected. This metallographic analysis is performed using standard metallographic techniques and specific equipment for this purpose. All fuel plates rejected are forwarded for uranium chemical recovery.

The metallurgical bonding quality of the assembled plates set, after rolling, is checked by means of bending tests. This test is performed at two occasions, after pre-cutting and after the final cut. This test is performed in the leftover material from the cutting operations. The material is extensively bent in an angle of 180° and in reverse. In case of bonding failure, which is easily detected by visual inspection, the fuel plate is rejected and sent to uranium chemical recovery.

6.3 Assembling of fuel elements

In IPEN, two types of fuel elements are manufactured. The standard fuel element consists of 18 fuel plates, 2 side plates (right and left), a nozzle, a handling pin and 8 screws. The control fuel element is composed of 12 fuel plates, two side plates (right and left), two guide plates, a nozzle, a dashpot and 12 screws. The dimensional characteristics of the fuel elements are specified. All structural components of the fuel element are manufactured according to designs that are part of the specifications.

The process begins with the assembling of fuel plates to form a case that is the structural body of the fuel element. The plates are fixed to the side plates (left and right) by mechanical clamping. Subsequently, the nozzle is fixed. For the standard fuel element, the handling pin is fixed on the side opposite to the nozzle. In the case of the control fuel element, the dashpot is fixed on the side opposite to the nozzle. After cleaning and inspection, the fuel element is packed and stored until transportation to the reactor. Figure 15 illustrates the steps for the fuel elements assembling process.

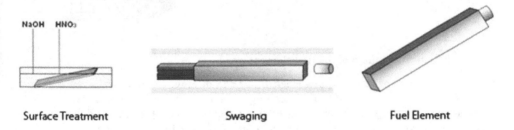

NaOH   HNO₃

Surface Treatment   Swaging   Fuel Element

Fig. 15. The process of assembling the fuel elements.

After fixing the fuel plates in the side plates to form the main case, the next component to be installed is the nozzle. The nozzle is used to fix the fuel elements in the reactor core. It is fixed by screws at the lower end of the main case. The nozzle is aligned with the case of fuel element through an adjustment operation by using precision measuring instruments. The holes in the nozzle that are used to fix the side plates are already machined. The holes to hold the external fuel plates at the nozzle are machined with the nozzle already fixed in the side plates, with the aid of a milling machine. The screws used are made with aluminum and are already qualified and properly cleaned before use. The final tightening is done after a previous dimensional characterization, once verified the alignment of the nozzle in the main case. If alignment does not meet the specification, it is adjusted. In the case of the control fuel element, the procedure for fixing the nozzle and dashpot is the same as described above.

The handling pin is used to handle the standard fuel element inside the reactor pool. It is installed at the upper end of the main case, which contains two holes where the handling pin is fixed by clinching. In this operation, the ends of the handling pin, which have cavities, are deformed by pressure with the aid of a drilling machine. In the case of the control fuel element, this pin is replaced by the dashpot, which is aimed at damping the control or security bars that operates within this type of fuel elements. Figure 16 illustrates the standard fuel element and its components.

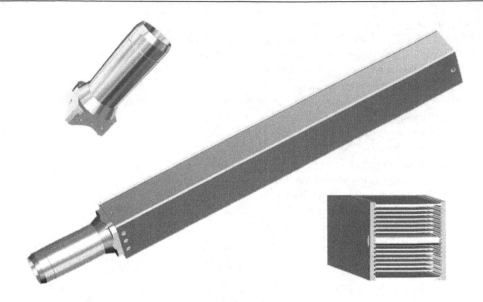

Fig. 16. Schematic illustration of the fuel element produced at IPEN.

Once qualified, the fuel element is washed in a bath of ethyl alcohol and dried manually with the aid of a jet of hot air. After this cleaning, a visual inspection is conducted, especially inside the cooling channels (the channels between the fuel plates), trying to detect possible obstructions caused by chips or foreign material. After washing and inspection, the fuel element is transferred to the reactor.

7. Uranium recovery and effluent treatment

A great variety of uranium residues must be recovered by chemicals means. A major source of such residues is uranium remaining in crucibles after melting and pouring. The recovery of solid or liquid uranium residues is vital because quantities are generated in every step of the process and this is a valuable material that must be recovered for reuse. Figure 17 displays a schematic diagram of the process showing the flow of products and residues.

The first step of the chemical recovery process is usually acid leaching to solubilize the uranium content. Any of several purification steps may then be employed to separate impurities such as iron, chromium, nickel, silicon, boron, etc. The end product of chemical recovery process is UF_4 which can be reduced to metal and then recycled. A typical sequence of chemical processing steps to recover uranium compounds from leach liquor is solvent extraction with tributyl phosphate, dinitration of purified uranyl nitrate solution to produce uranium trioxide (UO_3), and hydrogen reduction and hydrofluorination of UO_2 to UF_4. The technology of these operations is similar to that used in processing normal uranium.

Since chemical recovery will usually involve aqueous mixtures of uranium compounds, nuclear safety limits the critical dimensions of process equipment and imposes bath quantities within safe limits. If these factors are properly provided for chemical recovery unit design, the process operating costs will not be substantially raised by nuclear safety requirements.

The aggregate amount of scrap recycled via chemical recovery may reach 10% or more of finished fuel material weight. Chemical recovery is naturally more costly than direct recycle of metallic scrap to remelt. These considerations justify various expedients to by-pass chemical recovery by recycling metallic scrap. However, particular emphasis is given to the recovery of all residues solids and liquids because of the higher intrinsic value of the enriched material.

As an example, the IPEN process to produce U_3Si_2 involves metallic uranium as an intermediate product, through magnesiothermic reduction which produces slags containing uranium. The recovery process consists on slag lixivium of calcined by-products from metallic uranium reduction. The results from researching this process confirmed that this method could be integrated in treatment and recovery routines of uranium. The chemical route avoids dealing with metallic uranium since this material is unstable, pyroforic and extremely reactive. On the other hand, U_3O_8 is a stable oxide with low chemical reactivity, and it justifies the slags calcination of metallic uranium reduction by-products. This calcination occurs under oxidizing atmosphere and transforms the metallic uranium into U_3O_8. Some experiments have been carried out using diferente nitric molar concentrations, acid excess contents and temperature control of the lixivium process. The nitric lixivium main chemical reaction for calcined metallic uranium slags is represented by the equation:

$$U_3O_8 \text{ (s)} + 8\ HNO_3 \text{ (I)} \rightarrow 3\ UO_2(NO_3)_2 \text{ (I)} + 2\ NO_2(g) + 4H_2O(l) \tag{26}$$

The adopted process has the following parameters:

- Temperature and time: calcination of metallic uranium slag at 600°C during 3h;
- Granulometric control: sieving and segmentation of calcined slag in the range of 100-200 mesh;
- Concentration: lixivium adjustment of HNO_3 at 1 molar; HNO_3 excess (120%);
- Lixivium temperature: 40 - 50°C;
- Agitation: 300 rpm, turbine stem type (45° inclination).

As results, the full lixivium took 9 hours; the fluoride concentration in lixivium was 0,002g/L. Lixivium made at lower temperatures and lower nitric concentrations reduced both the magnesium and calcium fluorides solubility and the corrosion effect caused bifluoride ions was not prominent. This ensured a stable and secure lixivium from the operational point of view. The nitric dissolution of metallic uranium slags produced uranyl nitrate solution, which has been reused as a feed-in compound for uranium purification system made by solvent extraction method, using diluted n-tributhylphospate. The purified uranium product was then precipitated as ammonium diuranate (ADU) at 60°C, by injecting ammonium gas diluted with air. Aiming at returning the recovered product to the fuel fabrication cycle with nuclear quality level, the purified ADU was converted into uranium tetrafluoride (UF_4) by U_3O_8 route. The final yield in U content was 94%, proving the viability of IPEN´s slag recovering from uranium magnesiothermic reduction.

8. Acknowledgements

Thanks are due to IPEN for providing generously the technology of Nuclear Fuel Center, fully exemplified in this chapter, providing so nuclear know-how to a more peaceful, safer and healthy world. We are especially thankful to our colleagues who provided lots of information shown here, mainly Mr. Davilson Gomes da Silva who made many of the illustrations to qualify better this text.

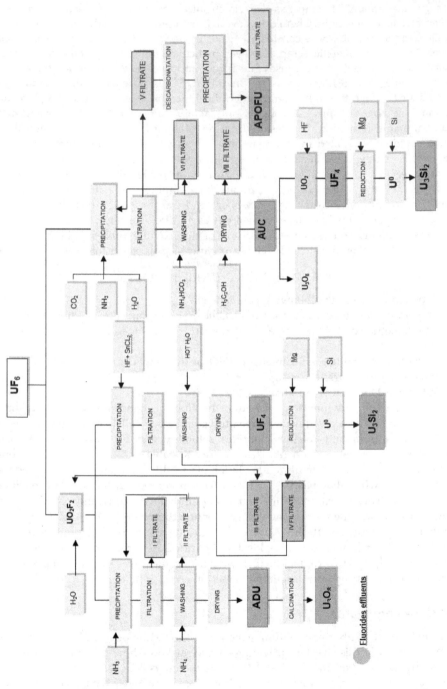

Fig. 17. Flowsheet MTR fuel processing (products and residues solid, liquids) (55; 54)

9. Conclusions

This chapter gave a general idea of the MTR fuel elements production for multipurpose and researching reactors that are producing radioisotopes throughout the world. Nowadays, the level of uranium enrichment is envisaged to be 20% (LEU), according to ruling requests of RERTR program. The given example of this production derived from IPEN/CNEN-São Paulo-Brazil, which produces through a well stablished routine to fabricate its own MTR fuel elements. Nevertheless, the technique to produce such elements has many variants, which are applied diversely from plant to plant.

As a final consideration, the future of fuel elements material, based on RERTR request, should also supply many high performance research reactors needing higher core densities of 6 to 9 gU/cm^3. This demand is not possible with U$_3$Si$_2$ elements, since its operational upper limit is less than 5 gU/cm^3. So, the presently envisaged product to reach this request is based on U-Mo alloy. Nevertheless, this product is not ready yet. Future prognosis are very confident that alloys U + 7 to 10wt%Mo should meet up this ability. This alloy production is still in experimental-pilot level, by this moment (2011), but with very consistent and pertinent results. For those willing to follow the development of this research, we indicate the transaction pages of RERTR and RRFM[2], where all papers and results are displayed freely.

10. References

[1] Cunningham, J. E. and Boyle, E. J. MTR-Type fuel elements. International Conference on Peaceful uses atomic energy. [ed.] United Nations. 1955, Vol. 9, pp. 203-7.

[2] Kaufman, A.R. *Nuclear reactor fuel elements, metallurgy and fabrication.* New York, NY, USA : Interscience, 1962.

[3] Holden, A.N. *Dispersions Fuel Elements.* New York, USA : Gordon & Breach, 1967.

[4] Cunningham, J.E., et al. Fuel Dispersions in Aluminium-Base Elements for research reactors. [ed.] Atomic Energy Comission. 1958, Vol. 1, pp. 269-97.

[5] Saller, H.A. Reaction Technology and Chemical Processing - Preparation, Properties and Cladding of Aluminum-Uranium Alloys. [ed.] United Nations. 1956, Vol. 9, pp. 214-20.

[6] Thurber, W.C. And Beaver, R.J. Segregation in Uranium-Aluminum Alloys and its Effect on the Fuel Loading of Aluminum-Base Fuel Element. [ed.] USAEC. 1958, pp. 9-29.

[7] Lennox, D.H. And Kelber, C.N. *Summary Report on the Hazards of the Argonaut Reactor.* Lemont, Mi : s.n., 1956. ANL – 5647.

[8] Kucera, W.J., Leitten, C.F. And Beaver, R.J. Specifications and Procedures Used in Manufacturing U3O8-Aluminium Dispersion Fuel Elements for Core I of the Puerto Rico Research Reactor. *Oak Ridge National Lab.* 1963.

[9] Knight, R.W., Binns, J. And Adamson Jr, G.M. Fabrication Procedures for Manufacturing High Flux Isotope Reactor Fuel Elements. *Oak Ridge National Lab.* Jun 1968.

[10] R.I., Beaver, Adamson Jr, G.M. And Patriarca, P. Procedures for Fabricating Aluminium-Base ATR Fuel Elements. *Oak Ridge National Lab.* Oak Ridge, Tenn.,, June, 1964.

---

[2]RERTR – Reduced Enrichment for Research and Test Reactors Program]: http://www.rertr.anl.gov/
RRFM – http://www.euronuclear.org/meetings/rrfm2011/transactions/RRFM2011-transactions.pdf

[11] Travelli, A. Current Status of the RERTR Program. *Development Fabrication and Application of Reduced-Enriched Fuels for Research and Test Reactor: Proceedings Held in Argonne.* 12-14 Nov 1980.

[12] J.L., snelgrove, Et Al. The Use of U3Si2 Dispersed in Aluminum in Plate-Type Fuel Elements for Research and Test Reactors. Oct 1987.

[13] Copeland, G.L., et al. Performance of Low-Enriched U3Si2-Al Dispersion Fuel Elements in the Oak Ridge Research Reactor. *RERTR.* Oct 1987.

[14] U.S.Regulatory Commission. Safety Evaluation Report related to the Evaluation of Low-Enriched Uranium Silicide-Aluminum Dispersion Fuel for Use in Non-Power Reactors. Jul 1988.

[15] Nazaré, S. New Low Enrichment Dispersion Fuels for Research Reactors Prepared by PM-Techniques. *J. Nucl. Mat.* 1984, Vol. 124, p. 14.

[16] —. Low Enrichment Dispersion Fuels for Research and Test Reactors. *Powder Met. Intern.* 1986, Vol. 18, 3, p. 150.

[17] Galkin, N.P., Et Al. *Technology of uranium.* Jerusalem : Israel Program for Scientific Translations, 1966. Cap.11.

[18] Snelgrove, J. L., Et Al. The use of U3Si2 dispersed in aluminum in plate-type fuel elements for research and test reactors. [ed.] October Argonne National Lab. Oct 1987.

[19] Van Wiencek, J. Uranium and fabrication. *Chemical Engineering Progress.* May 1954, p. 230.

[20] Seneda, J.A., Et Al. Recovery of uranium from the filtrate of ammonium diuranate prepared from uranium hexafluoride. *Journal of Alloys and Compounds.* 2001, Vols. 323-324, pp. 838-841.

[21] Instituto De Pesquisas Energéticas E Nucleares. *Implantação de um Centro de Processamento de Combustíveis no IPEN – Plano Diretor.* Ciclo do Combustível Nuclear. São Paulo : s.n., 1997. Relatório Interno MC.PT.0001.97.0.

[22] Cussiól, A.F. *Tecnologia para a preparação de tetrafluoreto de urânio. Fluoretação de UO2 obtido a partir de diuranato de amônio.* São Paulo, SP, Brasil : Escola Politécnica, Universidade de São Paulo, 1974. Dissertação de mestrado.

[23] Gisrgis, B.S. And Rofail, N.H. Reactivity of various UO3 modifications in the fluorination to UF4 by Freon 12. *J. of Nuclear Materials.* 1992, Vol. 195, p. 126.

[24] Gmelin, L. *Gmelins Handbuch der Anorganischen Chemie.* 8. Berlin : Springer-Verlag, 1980. C-8.

[25] Harrington, C.D. And Ruehle, E. *Uranium production technology.* . New Jersey : Van Nostrand, 1969.

[26] Katz, J. J. And Rabinowitch, E. *The chemistry of uranium, part 1 - The elements, its binary and related compounds.* New York : McGraw-Hill, 1951.

[27] Mellor, J.W. *A comprehensive treatise on inorganic and theorical chemistry.* London : Longmans, 1932. Vol. 12.

[28] Bolton, H.C. *Bull.Soc.Chim.* 1866, Vol. 6, 2, p. 450. apud Kat,J.J. & Rabinowitch, Z. The chemistry of uranium. Part I. The element, its binary and related compounds. New York, McGraw-Hill, 1951. p.355.

[29] —. *Z. Chem.* 1866, Vol. 2, 2, p. 353. apud KAT,J.J. & RABINOWITCH, Z. The chemistry of uranium. Part I. The element, its binary and related compounds. New York, McGraw-Hill, 1951. p.355..

[30] Alfredson, P.G. Australian experiments in the production of yellow cake and uranium fluorides. [ed.] International Atomic Energy Agency. *Production of yellow cake and uranium fluorides: proceedings of an Advisory Group meeting.* Jun 5-8, 1979, pp. 149-78.

[31] Allen, R.J., Petrow, H.G. And Magno, P.J. Precipitation of uranium tetrafluoride from aqueos solution by catalytic reduction. *Ind. Eng. Chem., 50(12).* 1958, Vol. 50, 12, pp. 1748-9.

[32] —. Preparation of sense, metal grade uranium tetrafluoride from uraniferous ores. [ed.] UNITED NATIONS. *Peaceful uses of atomic energy:proceedins of the 2nd international conference.* Sept. 1-13, 1958, Vol. 4.

[33] Gispert Benach, M., Et Al. Obtención de UF4 por reducción eletrolítica. I Estudios a escala de laboratorio. *Energ. Nucl.* 1972, Vol. 16, 80, pp. 623-36.

[34] Esteban Duque, A., Et Al. Producción de UF4 por reducción eletrolítica. II planta Piloto. *Energ. Nucl.* 1973, Vol. 17, 82, pp. 113-21.

[35] Scott, C.D., Adams, J.B. And Bresee, J.C. Fluorox process: production of UF6 in fluidized bed reactor. [ed.] Oak Ridge National Lab. 1960.

[36] Vogel, G.L., Et Al. Fluidized-bed techniques in producing uranium hexafluoride from ores concentrates. *Ind. Eng, Chem.* 1958, Vol. 50, 12, pp. 1744-7.

[37] Zidan, W. Elseaidy, I., , 26-29 October 1999, B. C.De Bariloche, Patagonia, Argentina. General Description and Production Lines Of The Egyptian Fuel Manufacturing Pilot Plant. *Proc.7th the Meeting of the International Group on Research ractors.* May 26-29, 1999.

[38] Opie, J.V. The preparation of pure uranium tetrafluoride by a wet process. . [ed.] Mallinckrodt Chemical Works. Apr 1, 1946.

[39] Ashbrook, A.W. And Smart, B.C. A review and update of refining pratice in Canada. In: International Atomic Agency. Production of yellow cake and uranium fluorides: proceedings od an Advisory Group meeting. Held in Paris, June 5-8, 1079. Vienna, 1980. p.261-8.

[40] Lenahan, K.J. Eldorado wet way process. In: Uranium 82: 12th annual hydrometallurgical meeting, held in Toronto, Aug. 29-Sept.1.1982.

[41] Harper, J. And Williams, A. E. *Factors influencing the magnesium reduction of uranium tetrafluoride.* In: Extraction and Refining of Rarer Metals, London: The Institute of Mining and Metallurgy, 1957, p. 143-162.

[42] Yemelyanov, V. S. E Yevstyukhin, A. I. *The metallurgy of nuclear fuel. London.* London : Pergamon Press, 1969.

[43] Huet, H. And Lorrain, C. Le procédé de magnésiothermie pour la préparation de l'uranium métallique. *Énergie Nucléaire.* 1967, Vol. 9, 3, pp. 181-188.

[44] Kubaschewski, O. High temperature reaction calorimetry - potencialities, limitations and application of results. *Thermochimica Acta.* 1978, Vol. 22, pp. 199-209.

[45] Beltran, A.D., Rivas Diaz, M. And Sanchez, A. F. Fabricacion de uranio metal. [ed.] Publicacion Bimestral de la Junta de Energia Nuclear. *Energia Nuclear.* mayo-junio 1972, pp. 295-320.

[46] Rand, M.H. And Kubaschewski, O. *The Thermochemical Properties of Uranium Compounds.* London : Oliver e Boyd, 1963.

[47] F., Fornarollo, Et Al. Recuperação de urânio em escórias geradas na produção de urânio metálico por magnesiotermia. *Revista Brasileira de Pesquisa e Desenvolvimento.* Novembro 2008, Vol. 10, 3, pp. 158-164.

[48] Saliba-Silva, A. M., Et Al. Fabrication of U3Si2 powder for fuels used in IEA-R1 nuclear research reactor. *Materials Science Forum*. IPEN, 2008, Vols. 591-93, pp. 194-199.

[49] Wiencek, T., Prokofiev, I. And Mcgann, D. Development and compatibility of Magnesium- Matrix Fuel Plates Clad With 6061 Aluminum Alloy", Proc. The International Meeting on Reduced Enrichment For research And Test Reactors. [ed.] ANL. *RERTR 1998*. Oct 18-23, 1998.

[50] Copeland, G. And Martin, M. Development of High - Uranium Loaded U3O8 - Al Fuel Plates. *Nuclear Technology*. 1982, Vol. 56.

[51] Tzou, G., Et Al. Analytical approach to the cold- and- hot bond rolling of sandwich sheet with outer hard and inner soft layers". *J. of Mater. Processing Technology*. Vols. 125-126, pp. 664-669.

[52] Frajndlich, E. U., Saliba-Silva, A. And Zorzetto, M. Xxi Rertr Meeting. Alternative Route For UF6 Conversion Towards UF4 to Produce metallic Uranium. *RERTR/DOE, 19°*. [Online] 1998. http://www.rertr.anl.gov/Fuels98/Elita.pdf.

[53] International Bio-Analytical Industries, Inc. Ammonium Uranyl Carbonate MSDS. [Online] [Cited: 05 16, 2011.]
     http://www.ibilabs.com/Ammonium%20Urany%20Carbonate%20MSDS.htm.

[54] Silva Neto, J. *Dry uranium tetrafluoride process preparation using the uranium hexafluoride reconversion process effluents*. São Paulo : CPG-IPEN/USP , 2008. MSc. Dissertation of University of São Paulo (in Portuguese).

[55] Frajndlich, E.U.C. *Chemical treatment study of an ammonium fluoride solution at the uranium reconversion plant*. São Paulo : CPG-IPEN/USP , 1992. MSc.Dissertation, University of São Paulo (in Portuguese).

4

Angular Dependence of Fluorescence X-Rays and Alignment of Vacancy State Induced by Radioisotopes

İbrahim Han
Ağrı İbrahim Çeçen University
Turkey

1. Introduction

This chapter concerns angular distribution measurements for fluorescence X-ray and the alignments of atoms with inner-shells vacancy resulting from ionization by radioisotope sources. The discussion on this topic is done by evaluating measurements of X-ray fluorescence parameters (such as cross-section, alignment parameter, polarization degree) from sample in various emission angles.

When an atom is ionized in one of its inner shells, the electrons rearrange themselves to fill the vacancy, with the transition energy released as a photon or transferred to another electron. The following X-ray or Auger electron may have an isotropic or non-isotropic angular distribution. The study of alignment of the inner-shell vacancy in ions can provide information about ionization process and the wave functions of inner-shell electrons, and calculations showed that the alignment was a sensitive testing parameter for theoretical models. For the last five decades there have been both theoretically and experimentally renewed efforts towards better understanding of the physics concerned with alignment of atoms with inner-shells vacancy and/or angular dependence of fluorescent X-rays emitted atoms induced photons or charged particle (electrons, protons, heavy ions). Generally, the alignments of atoms with inner-shells vacancy resulting from ionization by photons are investigated by measuring the anisotropic emission of X-ray lines using a detector (such as Si(Li) or Ge(Li)) and radioisotope photon source in various emission angles.

2. Historical background and current status of topic

The aim of paper interested in this topic is to determine the relationship between the angular distributions of X-rays with respect to total angular momentum values (J) of vacancy states. It is well-known that when radioisotope source, X-ray tube or charged particles produce vacancies in atoms at energy levels with $J>1/2$, the resulting ions will be aligned. The signature of this alignment is the anisotropic angular distribution of the emitted characteristic X-ray radiation, or the degree of polarization of the X-ray radiation. Total angular momentum (J) of vacancy states after photoionization is greater than 1/2, the population of its magnetic sub-states is non-statistical by the ionized atoms and this is reason of this anisotropic behavior. A lot of theoretical studies have been reported so far along this topic (Mehlhorn 1968; Mc Farlene, 1972; Berezhko and Kabachnik 1977; Sizov and Kabachnik, 1980, 1983) and the predictions of these researchers have been experimentally supported by some researchers (Schöler and Bell, 1978; Pálinkás, 1979, 1982; Wigger et al., 1984; Jesus et al., 1989; Mitra et al., 1996). The experimental study of alignment generally involves measurements of the angular distribution or polarization of the induced X-rays (Hardy et al., 1970; Döbelin et al., 1974; Jamison and Richard, 1977; Jistchin et al., 1979, 1983; Pálinkás, et al., 1981; Stachura et al., 1984; Bhalla, 1990; Mehlhorn, 1994; Papp 1999). In 1969, Cooper and Zare, (1969) first suggested a theoretical model relevant to aligned photon induced atoms. According to calculation by Cooper and Zare, (1969), after photoionization the inner-shell vacancy states have statistical population of magnetic substates. The vacancies produced after photoionizationin sub-shells are not be aligned at all and so the angular distribution of the fluorescent X-rays subsequent to photoionization will be isotropic. In 1972, 3 years after Cooper and Zare, the predictions of Flügge et al., (1972) showed that when vacancies are created in states with $J>1\ 2$, the population of its magnetic sub-states are non-statistical and therefore the resulting ions will be aligned. Mc Farlane (1972) calculated the polarization of X-rays from the decay of a vacancy in the $2p_{2/3}$ sub-shell using hydrogenic wave- functions in the Bethe approximation and the first Born approximation. After Caldwell and Zare (1977) first made an experimental investigation of the photon-induced alignment of Cd and they measured the degree of polarization of the emitted radiation from Cd. Since then, many experiments and calculations have been done to study the alignment of atoms and angular dependence characteristic X-rays by measuring either the angular distribution or the degree of polarization of the emitted X-rays. All these studies confirmed either alignment or not-alignment of the atoms after photoionization. The angular correlation between ionizing and fluorescent X-rays has been calculated relativistically, including all the radiation multipoles using single particle wavefunctions calculated in the Hartree–Slater model, by (Scofield, 1976). More recently, Scofield, (1989) used a relativistic model to study the angular distribution of the photoelectrons produced from photo- ionization by linear polarize photons and its inverse process (radiative recombination) in the energy region of 1–100keV. Scofield, (1989) found that the cross-section has a maximum at 90º compared to the direction of the incoming photons in the x–z plane (polarization plane) while the cross-section is independent of the angle between the incoming photon and the ejected electron in the y–z plane (normal to the polarization plane). Kamiya et al., (1979) measured L X-rays of Ho and Sm produced by protons and ^{3}He impacts with Si(Li) detector over the incident energy ranges $E_p = 0.75$–4.75MeV and $E_{3_{He}} = 1,5$–$9,4$ MeV in the direction of 90° to the projectile. Kamiya, et al., (1979) reported that the ratios of X-ray production cross-sections for the La and Ll lines depend clearly on projectile energy, but are independent of the projectile charge. Theoretical values of the alignment parameter for different states of various atoms calculated using the Herman-Skillman wave functions, have been reported by Berezhko and Kabachnik, (1977). The very strong anisotropy was reported for the emission of L lines for various elements by several scientists (Kahlon, et al., 1990a,b, 1991a,b; Ertuğrul, et al., 1995, 1996; Ertuğrul, 1996, 2002; Kumar, et al., 1999 Sharma and Allawadhi, 1999; Seven and Koçak, 2001,2002; Seven, 2004; Demir, et al., 2003). However, in all these investigations, the observed anisotropy is much higher than the predicted theoretical values of Scofield, (1976) and Berezhko and Kabachnik, (1977). On the other hand, anisotropic emission for L X-rays of Pb, Th and U was reported by some scientists (Mehta, et al., 1999; Kumar, et al 1999, 2001). Recently, Yamaoka et al., (2002, 2003) performed experiments using synchrotron radiation to determine the angular distribution of

L X-ray photons of Pb and Au. Although they found an isotropic distribution of the Pb L_3 lines within the experimental errors, non-isotropic angular distribution of the Au L_3 lines have been obtained. Papp and Campbell, (1992) reported the magnitude of the anisotropy and the alignment parameter for the L lines of Er. The alignment parameter of the ions of Xe was obtained by Küst, et al., (2003).

Kahlon et al., (1990a) reported experimental investigation of the alignment of the L_3 subshell vacancy state produced after photoionization in lead by 59.57 keV photons. The values of differential cross sections for the emission of the Ll, $L\alpha$, $L\beta$ and $L\gamma$ X-ray lines were determined at different emission angles varying from 40° to 120°. It was seen from the results that the Ll, and $L\alpha$ peaks show anisotropic emission, while the $L\beta$ and $L\gamma$ peaks are emitted isotropically. The angular dependence of emission intensity of L shell X rays induced by 59.57 keV photons in Pb and U was investigated by Kahlon et al., (1990b) measuring the normalized intensities of the resolved L X-ray peaks at different angles varying from 40° to 140°. It was observed that while the Ll and $L\alpha$ peaks (originating from J=3/2 state) show some anisotropic angular distribution, the emission of the $L\beta$ and $L\gamma$ peaks are emitted isotropically. Kahlon et al., (1991a) measured the angular distribution and polarization of the L shell fluorescent X-rays excited by 59.54 keV photons in Th and U. It was found that the $L\gamma$ group of L X-rays is isotropic in spatial distribution and unpolarized but, the Ll and $L\alpha$ groups are anisotropically distributed and polarized. Although no anisotropy of the $L\beta$ group is detected, it was slightly polarized. Kahlon et al., (1991b) investigated the differential cross sections for emission of Ll, $L\alpha_2$, $L\alpha$, $L\beta$ and $L\gamma$ groups of L X-ray lines induced in Au by 59.54 keV photons at different angles varying from 40° to 120°. The L X-rays represented by Ll, $L\alpha_2$ and $L\alpha$ peaks were found to be anisotropic in the spatial distribution while those in $L\beta$ and $L\gamma$ peaks were isotropic. Papp and Campbell, (1992) measured angular distributions of the L_l, $L\alpha_{1,2}$ and $L\beta_{2,15}$ transitions of erbium in the angular range of 70°–150° following photoionization by 8.904 keV photons. A Johansson-type monochromatic was used to select the Cu $K\beta_1$ line for ionization. Anisotropy parameters for Ll, $L\alpha_{1,2}$ and $L\beta_{2,15}$ were found as 0.052±0.016, 0.16±0.022 and 0.012±0.015, respectively. Ertugrul et al., (1995, 1996a, 1996b) measured differential cross-sections for the emission of Ll, $L\alpha$, $L\beta$ and $L\gamma$ X-rays of Au, Hg, Tl, Pb, Bi, Tb and U at different emission angles varying from 45° to 135°. They found that Ll and $L\alpha$ peaks are emitted isotropically, while $L\beta$ and $L\gamma$ peaks show anisotropic emission. Sharma and Allawadhi, (1999) measured values of Ll, $L\alpha$ and $L\beta$ differential X-ray production cross sections in Th and U at 16.896 and 17.781 keV at emission angles 60°, 70°, 80° and 90°. From the results of the measurements it was evident that, in the present case, all the three Ll, $L\alpha$ and $L\beta$ differential X-ray production cross sections depend on the emission angle and thus, the emission is anisotropic. Demir et al., (2000) indicated differential cross-sections for the emission of M shell fluorescence X-rays from Pt, Au and Hg by 5.96 keV photons at seven angles ranging from 50° to 110° at. The differential cross-sections were found to decrease with increase in the emission angle, showing an anisotropic spatial distribution of M shell fluorescence X-rays. Seven and Koçak (2001, 2002) measured the Ll, $L\alpha$, $L\beta$ and $L\gamma$ X-ray production cross-sections in U, Th, Bi, Pb, Tl, Hg, Au, Pt, Re,W, Ta, Hf, Lu, and Yb using 59.5 keV incident photon energies in the angular range 40°-130°. Although differential cross sections for $L\beta$ and $L\gamma$ X-rays were found to be angle independent within experimental error, those for the Ll and $L\alpha$ X-rays were found to be angle dependent. Ertugrul et al., (2002) measured the alignment parameter the $I_{L\alpha}/I_{Ll}$ intensity ratio. The Ll and $L\alpha$ X-rays of the elements were measured with a Si (Li) detector at a direction of 90° to the projectile. The L3 edges of Nd,

Gd, Tb, Dy, Ho, Er, Yb, Hf, Ta, W, Au, Hg, Tl, Pb, Bi, Th and U the elements were excited with the K X-ray energy of 17.781(MoK$_{\alpha,\beta}$), 16.896(NbK$_{\alpha,\beta}$), 14.980(RbK$_\beta$), 13.300(BrK$_\beta$), 12.503(SeK$_\beta$), 12.158(BrK$_{\alpha,\beta}$), 10.983(GeK$_\beta$), 10.073(GeK$\alpha_{1,\beta}$), 9,572(ZnK$_\beta$), 8.976(CuK$_{\beta2}$), 8.907(CuK$_\beta$), 8.265(NiK$_\beta$), 7.649(CoK$_{\beta1}$), 6.490(MnK$_{\beta1}$) keV from the selected elements, respectively. They noticed that the L$_3$ X-rays show large anisotropy, the measured alignment parameter varying from -0.115 to +0.355. Demir et al., (2003) reported Ll, $L\alpha$, $L\beta$ and $L\gamma$ X-ray differential cross-sections, fluorescence cross-sections and σ_{L1}, σ_{L2} and σ_{L3} subshell fluorescence cross-sections for Er, Ta, W, Au, Hg and Tl at an excitation energy of 59.6 keV. The differential cross-sections for these elements have been measured at different angles varying from 54° to 153°. The Ll and $L\alpha$ groups in the L X-ray lines were found to be spatially anisotropic, while those in the $L\beta$ and $L\gamma$ peaks are isotropic. The Ll, $L\alpha$, $L\beta_{2,4}$, $L\beta_{1,3}$ and $L\gamma$ X-ray production cross-sections and L-subshell fluorescence yields ω_1 and ω_2 in Th and U have been determined by Seven (2004) at an incident photon energy of 59.54 keV by measuring differential cross-sections with angles changing from 40° to 130°. The Ll, $L\alpha$ and $L\beta_{2,4}$ X-rays have an anisotropic spatial distribution while $L\beta_{1,3}$ and $L\gamma$ X-rays have isotropic spatial distributions. Özdemir et al., (2005) measured the angular dependence of L$_3$ subshell to M-shell vacancy transfer probabilities for the elements Lu, Hf, Ta, W, Os and Pt at the excitation energies of 5.96 keV and K X-rays of Zn, Ga, Ge, and As, respectively, at seven angles varying from 120° to 150°. It was observed that angular dependence from L$_3$ subshell to M-shell vacancy transfer probabilities increase with increasing cosθ. The angular dependence of M X-ray production differential cross-sections for selected heavy elements between Lu and Pt have been measured by Durak (2006) at 5.59 keV of incident photon energy and at seven emission angles in the range of 120°-150°. Angular dependence of M X-ray production differential cross sections has been derived, using the M-shell fluorescence yields, experimental total M X-ray production cross sections and theoretical M-shell photoionization cross sections. M X-ray production differential cross-sections were found to decrease with increase in the emission angle, showing an anisotropic spatial distribution of M X-rays. Angular dependence from L$_3$ subshell to M-shell vacancy transfer probabilities for selected heavy elements from Au to U were measured by Özdemir and Durak (2008) at different angles varying from 120° to 150°. It was observed that angular dependence from L$_3$-subshell to M-shell vacancy transfer probabilities increase with increasing cosθ. Apaydın et al., (2008) measured Mi ($i = a + \beta$) X-ray production differential cross sections for Re, Bi and U elements at the 5.96 keV incident photon energy in an angular range 135°-155°. They found that the angular dependence M X-rays production cross sections decrease with increase in the emission angle, showing anisotropic spatial distribution

Kumar et al., (1999) investigated the angular dependence of emission of L x-rays following photoionization at 22.6 and 59.5 keV in 82Pb by measuring the intensity ratios $I_{Ll}/I_{L\gamma}$, $I_{L\alpha}/I_{L\gamma}$ and $I_{L\beta}/I_{L\gamma}$ at different angles varying from 50° to 140°. The measured intensity ratios for various L x-rays were found to be angle independent within experimental error. Mehta et al., (1999) measured the L_l, L_α, L_η, $L\beta_6$, $L\beta_{2,4}$, $L\beta_{1,3}$, $L\beta_{9,10}$ and L_γ x-ray production differential cross sections in 92U using the 22.6- and 59.5-keV incident photon energies in an angular range 43°–140°. Differential cross sections for various L x rays were found to be angle independent within experimental error. Puri et al., (1999) measured The Ll, $L\alpha$, $L\beta_{2,4}$, $L\beta_{1,3}$ and $L\gamma_{1,5}$ X-ray production differential cross sections in 90Th have at 22.6 keV incident photon energy in an angular range 50° -130° The measured differential cross sections for various L X-rays were found to be angle-independent within experimental error. Kumar et al., (2001a) measured the the Ll, $L\alpha$ and $L\beta_{2,5,6,715}$ X-ray fluorescence (XRF) differential cross-sections in Pb at the 13.6 keV incident photon energy ($E_{L3} < E_{inc} < E_{L2}$, E_{Li} being the Li sub-shell binding energy) and in the angular range 90-160°. At this incident photon energy, the L_3 sub-shell vacancies ($J = 3/2$) are produced only due to the direct ionization and the reduction in the observed anisotropy in the emission of the Ll, $L\alpha$ and $L\beta_{2,5,6,715}$ X-rays due to the transfer of unaligned L_1 and L_2 subshell vacancies ($J = 1/2$) to the L_3 sub-shell through Coster–Kronig transitions was eliminated. The differential cross-sections for various x-rays were found to be angle-independent within experimental error. The L X-ray production (XRP) differential cross sections in Th and U have been measured by Kumar et al., (2001b) at the 17.8 keV incident photon energy ($E_{L3} < E_{inc} < E_{L2}$, E_{Li} is the Li subshell ionization threshold) in an angular range 90°-160°and at the 25.8 and 46.9 keV incident photon energies ($E_{L1} < E_{inc} < E_K$) at an angle of 130°. The present measurements rule out the possibility of a strong angular dependence of differential cross sections for various L_3 subshell X-rays following selective photoionization of the L_3 subshell. Tartari et al., (2003) investigated the anisotropy of L X-ray fluorescence induced by 59.54 keV unpolarized photons by means of an experimental procedure which allows the relative L X-ray production cross section to be evaluated without taking account of the angular set-up and the instrumental efficiency. Thick targets of Yb, Hf, Ta, W and Pb are considered, and the angular trend of the relative experimental ratios, $I_{L\alpha}/I_{L\beta}$, is calculated by simple evaluations of the peak area alone. Within the experimental uncertainties, which were found to be of the order of 1.6% in the worst cases, the results do not show any significant angular dependence of the $L\alpha$ emission lines. Santra et al., (2007) measured the angular distribution of the L X-ray fluorescent lines from Au and U induced by 22.6-keV X-rays in the angular range of 70°–150°. No strong anisotropy was observed as mentioned by some groups. In the case of Au, a maximum anisotropy of 5% was observed while for U it was within experimental errors 2%. From the angular distribution of the L_l line of Au, the alignment parameter was obtained and its value was found to be 0.10±0.14. Kumar et al., (2008) investigated alignment of the $M_3(J= 3/2)$, $M_4(J= 3/2)$ and $M_5(J= 5/2)$ subshell vacancy states produced following photoionization in the M_i (i=1-5) subshells of Au, Bi, Th and U through angular distribution of the subsequently emitted M X-rays. The unpolarized Mn K X rays (E_{KX}=5.97 keV) from the 55Fe radioisotope were used to ionize the Mi subshells in an angular range 90°-160°and the emitted M X-rays were measured under vacuum using a low energy Ge detector. The M X-ray spectra taken at different emission angles were normalized using the isotropically emitted K shell ($J= 1/2$) X-rays measured simultaneously from a 23V thin target placed adjoining the M X-ray target. The present precision measurements infer that anisotropy in the $M_{\alpha\beta\gamma}$ X-ray emission shows trends and order of magnitude predicted by theoretical calculations, i.e., anisotropy parameter (β_2)~0.01.

In the recent experimental study (Han et al., 2008), the angular distribution of characteristic K and L X-rays, emitted from Sm, Eu, Gd Tb, Dy, Ho, and Er as a result of K and L shell vacancies produced by 59.54 keV photon impact was investigated. Thus, K and L X-rays emitted from these elements were simultaneously measured in the same experimental geometry. In this study, Sm, Eu, Gd, Tb, Dy, Ho, and Er lanthanides were chosen since both K shell and L shell electrons of these elements can be excited simultaneously by an Am-241 point source. Also, K and L peaks of the chosen elements are well resolved. Earlier experimental investigations have been only performed on the K X-ray cross sections or on the angular distribution of L X-rays. This is the first report of the angular distributions of Li X-ray and Ki X-ray ($i = a, \beta$) cross sections for Sm, Eu, Gd, Tb, Dy, Ho, and Er at different angles. It is well known that K X-ray cross sections have no angular dependency ($J = 1/2$). The experimental investigation on K X-ray cross sections at different angles was made to check the validity of the angular dependency of experimental L X-ray cross sections. The experimental K X-ray cross sections were compared with theoretically calculated values and fairly good correspondence was observed. This means that the present measurements regarding angular dependency of L X-rays are reliable.

In following the work of us (Han et al., 2009) experimental results of the angular distribution of characteristic X-rays were introduced. We preferred to use of $I_{La}/I_{Ll}(\theta)$ intensity ratios to obtain the values of alignment parameters (A_2). In that case, the background subtraction problem is considerably reduced and statistical errors are significantly less. It was observed from measured intensities that La and Ll X-ray intensities for the L_3 sub-state depended on the emission angle, meaning that La and Ll X-rays had an anisotropic spatial distribution. Thus, the La to Ll intensity ratios for a set of elements was determined and alignment parameters for each element were obtained using these ratios. In this study, three L sub-shells electrons were excited. Therefore, alignment parameter values are influenced by Coster–Kronig transitions from vacancies induced in the L_1 or L_2 sub-shells. L_1 and L_2 sub-shells have the same J= 1/2 value therefore the transferred vacancies are not-aligned and the observed anisotropy of the X-rays is attenuated. For this reason, corrected value of the alignment parameter was calculated using attenuation factor F. If photon energies exciting only L_3 sub-shell electrons are chosen, the alignment parameter will be independent from Coster–Kronig transitions

In more recently study (Han and Demir, 2011a), we investigated the angular distribution of characteristic L X-rays emitted from heavy elements (Pt, Au, Pb, Bi, Th and U) as a result of L shell vacancy production by 59.54 keV photon impact and angular distribution of Compton scattering photons from the same elements. Thus, emitted fluorescent L X-rays and Compton scattering photons from elements were simultaneously measured in the same experimental geometry. Earlier experimental investigations have been only performed on the angular distribution of L X-rays or Compton scattering photons. This is the first report of the angular distribution of Li ($i=$ l, a, β and γ) X-rays fluorescent and Compton scattering differential cross sections for Pt, Au, Pb, Bi, Th and U at different angles in the same experimental geometry. It is well known that Compton scattering differential cross sections have angular distribution. The experimental investigation on Compton scattering differential cross sections at different angles was made to check the validity of angular distribution of experimental L X-rays fluorescent differential cross sections. The experimental Compton scattering differential cross sections were compared with theoretically calculated values and fairly good correspondence was observed. This means that the present measurements regarding angular distribution of L X-rays are reliable. In the meantime, L_3-subshell alignment of Th and U ionized by 59.5 keV photons has been investigated by evaluating the angular dependence of Li ($i=l$, a, η, β and γ) X-ray lines. The angular dependence measurements were performed by measuring the fluorescence cross section, σ_{Li} ($i=$ l, a, η, β and γ) and $\sigma_{Ll}/\sigma_{L\gamma}$, $\sigma_{L\eta}/\sigma_{L\gamma}$, $\sigma_{La}/\sigma_{L\gamma}$ and $\sigma_{L\beta}/\sigma_{L\gamma}$ ratios at different angles. It was observed from the measurements that Li ($i=l$ and a) X rays for the L_3-subshell depended on the emission angle and had an anisotropic spatial distribution. On the other hand, there was no dependence of emission angle and any significant anisotropy for other L X rays. The both Ll and La X-rays originate from the filling of vacancies in states L_3-subshell with J = 3/2. The results of measurements indicate that the L_3-subshell vacancy states with J =3/2 are aligned, whereas L_1, and L_2 vacancy states with J =1/2 are non-aligned. Integral cross-sections for the Li ($i=$ l, a, η, β and γ) X-rays and L subshell fluorescence yields ω_i ($i=$ 1, 2 and 3) were also determined and results were compared with theoretically calculated values and results of others and fairly good correspondence was observed. The $L\gamma$ X-rays, originating purely from the L_1 and L_2 subshells, having isotropic emission were used to normalize the intensities of the anisotropic Ll and the $L\alpha$ X-rays originating from the L_3 subshell. It was observed from measurements that Ll and $L\alpha$ X-ray for the L_3 sub-state depended on the emission angle, meaning that Ll and $L\alpha$ X-rays had an anisotropic spatial distribution. On the other hand, the $L\beta$ and $L\gamma$ X-rays don't show any significant anisotropy. The fluorescence cross sections for Ll and $L\alpha$ X-rays are decreased with increased emission angles (Han and Demir, 2011b).

3. Conclusion

In the light of all these, above; data from different researchers show contradictory and the existing results on the angular dependence of fluorescence X-ray and the alignment of atoms with inner-shells vacancy following ionization are still controversial and quite confusing. Therefore, more experimental and theoretical investigations should be required to settle the present discrepancies

4. Acknowledgment

I thank to M.R. Kacal for his help and advice during the preparation of this chapter.

5. References

Apaydın, G., Tırasoglu, E., Sogut, O., 2008. Measurement of angular dependence of M X-ray production cross-sections in Re, Bi and U at 5.96 keV Eur. Phys. J. D 46, 487–492

Berezhko, E.G., Kabachnik, N.M., 1977. Theoretical study of inner-shell alignment of atoms in electron impact ionisation: angular distribution and polarization of X-rays and Auger electrons. J. Phys. B At. Mol. Opt. Phys. 10, 2467–2477.

Bhalla, C.P., 1990. Angular distribution of Auger electrons and photons in resonant transfer and excitation in collisions of ions with light targets. Phys. Rev. Lett. 64, 1103–1106.

Caldwell, C.D., Zare, R.N., 1977.Alignment of Cd atoms by photoionization. Phys. Rev. A 16, 255–262.

Cooper, J., Zare, N., 1969. Potoelectron angular distributions. In:Geltman, S., Mahanthappa, K.T., Brittin, W.E.(Eds.), Lecturesin Theoretical Physics: Atomic Collision Processes, vol. X1C. Gordon and Breach, NewYork, pp. 317–337.

Demir, L., Şahin, M., Kurucu, Y., Karabulut, A., Şahin, Y., 2000. Measurement of angular dependence of photon-induced differential cross-sections of M X-rays from Pt, Au and Hg at 5.96 keV. Radiat. Phys. Chem. 59 355-359

Demir, L., Şahin, M., Kurucu, Y., Karabulut, A., Şahin, Y., 2003. Angular dependence of Ll, $L\alpha$, $L\beta$ and $L\gamma$ X-ray differential and fluorescence cross-sections for Er, Ta, W, Au, Hg and Tl Radiat. Phys. Chem. 67, 605–612

Döbelin, E., Sandner, W., Mehlhorn, W., 1974. Experimental study of inner shell alignment of atoms in electron impact ionization. Phys. Lett. A 49, 7–8.

Durak, R., 2006. Measurement of angular dependence of M X-ray production differential cross-sections in heavy elements at 5.96 keV. Can. J Anal. Sci. Spectrosc. 51, No. 2.

Ertuğrul, M., Büyükkasap, E., Erdoğan, H., 1996a. Experimental investigation of the angular dependence of photon-induced differential cross-sections of L X-rays from U, Th and Bi at 59.5 keV. Il Nuovo Cimento D, 18, 671–676.

Ertuğrul, M., Büyükkasap, E., Küçükönder, A., Kopya, A.İ., Erdoğan, H., 1995. Anisotropy of L -shell X-rays in Au and Hg excited by 59.5 keV photons. Il Nuovo Cimento D, 17, 993-998.

Ertuğrul, M., Öz, E., Şahin, Y., 2002. Measurement of alignment parameters for photon induced L_3 vacancies in the elements $59 \leq Z \leq 92$. Physica Scripta 66, 289–292.

Ertuğrul, M.,1996b. Measurement of cross-sections and Coster–Kronig transition effect on L sub-shell X-rays of some heavy elements in the atomic range $79 \leq Z \leq 92$ at 59.5 keV. Nucl. Instrum. Methods Phys. Res. B 119, 345–351.

Flügge, S.,Mehlhorn, W.,Schmidt,V.,1972. Angular distribution of auger electrons following photoionization. Phys.Rev.Lett. 29, 7–9 Erratum: Phys.Rev.Lett. 29, 1288.

Han, I., Demir, L., 2011a. Angular distribution of fluorescent L X-rays and Compton scattering photons Spectrosc. Lett. 44, 95–102.

Han, I., Demir, L., 2011b. Angular dependence of L_3-subshell X-ray emission following photoionisation. J X-Ray Sci. Technol. 19, 13–21.

Han, I., Sahin, M., Demir, L., 2008. Angular variations of K and L X-ray fluorescence cross sections for some lanthanides Can. J. Phys. 86, 361–367.

Han, I., Sahin, M., Demir, L., 2009. The polarization of X-rays and magnetic photoionization cross-sections for L_3 sub-shell. Appl. Radiat. Isot. 67, 1027–1032.

Hardy, J., Henins, A., Bearden, J.A., 1970. Polarization of the L_{a1} X-rays of mercury. Phys. Rev. A 2, 1708–1710.

Jamison, K.A., Richard, P., 1977. Polarization of target K X-rays. Phys. Rev. Lett. 38, 484–487.

Jesus, A.P., Ribeiro, J.P., Niza, I.B., Lopes, J.S., 1989. L_3-subshell alignment of Au induced by proton, deuteron and alpha-particle impact. J. Phys. B At. Mol. Opt. Phys. 22, 65.

Jitschin, W., Hippler, R., Shanker, R., Kleinpoppen, H., Schuch, R., Lutz, H.O., 1983. L X-ray anisotropy and L_3-sub-shell alignment of heavy atoms induced by ion impact. J. Phys. B At. Mol. Opt. Phys. 16, 1417–1431.

Jitschin, W., Kleinpoppen, H., Hippler, R., Lutz, H.O., 1979. L-shell alignment of heavy atoms induced by proton impact ionization. J. Phys. B At. Mol. Opt. Phys. 12, 4077–4084.

Kahlon, K.S., Aulakh, H.S., Singh, N., Mittal, R., Allawadhi, K.L., Sood, B.S., 1990a. Experimental investigation of alignment of the L_3 sub-shell vacancy state produced after photoionization in lead by 59.57 keV photons. J. Phys. B At. Mol. Opt. Phys. 23, 2733-2743.

Kahlon, K.S., Aulakh, H.S., Singh, N., Mittal, R., Allawadhi, K.L., Sood, B.S., 1991a. Measurement of angular distribution and polarization of photon-induced fluorescent x rays in thorium and uranium. Phys. Rev. A 43, 1455–1460.

Kahlon, K.S., Shatendra, K., Allawadhi, K.L., Sood, B.S., 1990b. Experimental investigation of angular dependence of photon induced L shell X-ray emission intensity. Pramana 35, 105–114.

Kahlon, K.S., Singh, N., Mittal, R., Allawadhi, K.L., Sood, B.S., 1991b. L_3-sub-shell vacancy state alignment in photon–atom collisions. Phys. Rev. A 44, 4379–4385.

Kamiya,M., Kinefuchi, Y., Endo, H., Kuwako, A., Ishii, K., Morita, S., 1979. Projectile- energy dependence of intensity ratio of La to Ll X-rays produced by proton and ^{3}He impacts on Ho and Sm. Phys. Rev. A 20, 1820–1827.

Kumar, A., Garg, M.L., Puri, S., Mehta, D., Singh, N., 2001a. Angular dependence of L_3 x-ray emission following L_3 sub-shell photoionization in Pb. X-Ray Spectrom. 30, 287–291

Kumar, A., Puri, S., Mehta, D., Garg, M.L., Singh, N., 1999. Angular dependence of L x-ray emission in Pb following photoionization at 22.6 and 59.5 keV. J. Phys. B At. Mol. Opt. Phys. 32, 3701–3709.

Kumar, A., Puri, S., Shahi, J.S., Garg, M.L., Mehta, D., Singh, N., 2001b. L X-ray productioncross-sections in Th and U at 17.8, 25.8 and 46.9 keV photon energies. J. Phys. B At. Mol. Opt. Phys. 34, 613–623.

Kumar, S., Sharma, V., Mehta, D., Singh, N., 2008. Alignment of Mi (i=3–5) subshell vacancy states in 79Au, 83Bi, 90Th, and 92U following photoionization by unpolarized Mn K x rays. Phys. Rev. A 77, 032510

Küst, H., Kleiman, U., Mehlhorn, W., 2003. Alignment after Xe L_3 photoionization by synchrotron radiation. J. Phys. B At. Mol. Opt. Phys. 36, 2073.

Mc Farlane, S.C., 1972. The polarization of characteristic X radiation excited by electron impact. J. Phys. B At. Mol. Opt. Phys. 5, 1906–1915.

Mehlhorn, W., 1968. On the polarization of characteristic X radiation. Phys. Lett. A 26, 166–167.

Mehlhorn,W., 1994. Alignment after inner-shell ionization by electron impact near and at threshold. Nucl. Instrum. Methods Phys. Res. B. 8, 7227–7233.

Mehta, D., Puri, S., Singh, N., Garg, M.L., Trehan, P.N., 1999. Angular dependence of L X-ray production cross-sections in U at 22.6 and 59.5 keV photon energies. Phys. Rev. A 59, 2723–2731.

Mitra, D., Sarkar, M., Bhattacharya, D., Chatterjee, M.B., Sen, P., Kuri, G., Mahapatra, D.P., Lapicki, G., 1996. L_3-subshell alignment in gold and bismuth induced by low-velocity carbon ions. Phys. Rev. A 53, 2309–2313.

Ozdemir, Y., Durak R., 2008. Angular dependence from L_3-subshell to M-shell vacancy transfer probabilities for heavy elements using EDXRF technique. Annals of Nuclear Energy 35 1335–1339.

Ozdemir, Y.,Durak, R., Esmer, K., Ertugrul, M., 2005. Measurement of angular dependence from L_3-subshell to M-shell vacancy transfer probabilities for the elements in the atomic region 71≤Z≤78. J Quant. Spectrosc. Radiat. Transf. 90 161–168.

Pálinkás, J., Sarkadi, L., Schlenk, B., Török, I., Kálmán, Gy., 1982. L_3-subshell alignment of gold by C$^+$ and N$^+$ impact ionisation. J. Phys. B At. Mol. Opt. Phys. 15, L451.

Pálinkás, J., Schlenk, B., Valek, A., 1979. Experimental investigation of the angular distribution of characteristic X-radiation following electron impact ionisation. J. Phys. B At. Mol. Opt. Phys. 12, 3273.

Pálinkás, J., Schlenk, B., Valek, A., 1981. The Coulomb deflection effect on the L_3-sub-shell alignment in low-velocity proton impact ionization. J. Phys. B At. Mol. Opt. Phys. 14, 1157–1159.

Papp, T., 1999. On the angular distribution of X-rays of multiply ionized atoms. Nucl. Instrum. Methods Phys. Res. B 154, 300–306.

Papp, T., Campbell, J.L., 1992. Non-statistical population of magnetic substates of the erbium L_3 subshell in photoionization. J. Phys. B At. Mol. Opt. Phys. 25, 3765.

Puri, S., Mehta, D., Shahid, J.S., Garg, M.L., Singh, N., Trehan, P.N., 1999. Photon-induced L X-ray production di€erential cross sections in thorium at 22.6 keV. Nucl. Instrum. Methods Phys. Res. B 152 19-26

Santra, S., Mitra, D., Sarkar, M., Bhattacharya, D., 2007. Angular distribution of Au and U L x rays induced by 22.6-keV photons. Phys. Rev. A 75, 022901.

Schöler, A., Bell, F., 1978. Angular distribution and polarization fraction of characteristic X-radiation after proton impact. Z. Phys. A 286, 163–168.

Scofield, J.H., 1976. Angular dependence of fluorescent X-rays. Phys. Rev. A 14, 1418–1420.

Scofield, J.H., 1989. Angular and polarization correlations in photoionization and radiative recombination. Phys. Rev. A 40, 3054–3060.

Seven, S., 2004. Measurement of angular distribution of fluorescent X-rays and L subshell fluorescence yields in thorium and uranium. Radiat. Phys. Chem. 69, 451–460

Seven, S., Koçak, K., 2001. Angular dependence of L x-ray production cross sections in seven elements from Yb to Pt at a photon energy of 59.5 keV. J. Phys. B At. Mol. Opt. Phys. 34, 202.

Seven, S., Koçak, K., 2002. Angular dependence of L x-ray production cross-section in seven elements from Au to U at 59.5 keV photon energy. X-Ray Spectrom. 31, 75–83.

Sharma, J.K., Allawadhi, K.L., 1999. Angular distribution of Lβ X-rays from decay of L$_3$ sub-shell vacancies in uranium and thorium following photoionization. J. Phys. B At. Mol. Opt. Phys. 32, 2343–2349.

Sizov, V.V., and Kabachnik, N.M., 1980. Inner-shell alignment of atoms in ion-atom collisions. I. Impact ionisation. J. Phys. B At. Mol. Opt. Phys. 13, 1601.

Sizov, V.V., and Kabachnik, N.M., 1983. Inner-shell alignment of atoms in ion-atom collisions. III. Light target atoms. J. Phys. B At. Mol. Opt. Phys. 16, 1565.

Stachura, Z., Bosch, F., Hambsch, F.J., Liu, B., Maor, D., Mokler, P.H., Schonfeldt, W.A., Wahl, H., Cleff, B., Brussermann, M., Wigger,J., 1984. Anisotropy of Ll X-ray transition observed in 1.4 MeV N^{-1} heavy ion-atom collisions. J. Phys. B At. Mol. Opt. Phys. 17, 835–847.

Tartari, A., Baraldi, C., Casnati, E., Re, A.D., Fernandez, J.E., Simone, T., 2003. On the angular dependence of L x-ray production cross sections following photoionization at an energy of 59.54 keV. J. Phys. B: At. Mol. Opt. Phys. 36 843–851.

Wigger, J., Altevogt, H., Brüssermann, M., Richter, G., Cleff, B., 1984. M$_3$, M$_4$ and M$_5$ alignment of thorium by proton impact ionisation. J. Phys. B At. Mol. Opt. Phys. 17, 4721.

5

Determination of Chemical State and External Magnetic Field Effect on the Energy Shifts and X-Ray Intensity Ratios of Yttrium and Its Compounds

Sevil Porikli[1] and Yakup Kurucu[2]
Erzincan University, Faculty of Art and Sciences, Department of Physics
Atatürk University, Faculty of Sciences, Department of Physics
Turkey

1. Introduction

The term 'X-ray fluorescence analysis' (XRF) refers to the measurement of characteristic fluorescent emission resulting from the deexcitation of inner shell vacancies produced in the sample by means of a suitable source of radiation. For a particular energy (wavelength) of fluorescent light emitted by a sample, the number of photons per unit time (generally referred to as peak intensity or count rate) is related to the amount of that analyte in the sample. The counting rates for all detectable elements within a sample are usually calculated by counting, for a set amount of time, the number of photons that are detected for the various analytes' characteristic X-ray energy lines. It is important to note that these fluorescent lines are actually observed as peaks with a semi-Gaussian distribution because of the imperfect resolution of modern detector technology. Therefore, by determining the energy of the X-ray peaks in a sample's spectrum, and by calculating the count rate of the various elemental peaks, it is possible to qualitatively establish the elemental composition of the samples and to quantitatively measure the concentration of these elements.

XRF is an analytical method to determine the chemical composition of all kinds of materials. The materials can be in solid, liquid, powder, filtered or other form. XRF can also sometimes be used to determine the thickness and composition of layers and coatings. The method is fast, accurate and non-destructive, and usually requires only a minimum of sample preparation. Applications are very broad and include the metal, cement, oil, polymer, plastic and food industries, along with mining, mineralogy and geology, and environmental analysis is of water and waste materials. XRF is also a very useful analysis technique for research and pharmacy.

For routine XRF analysis, two major approaches are distinguishable based on the type of detector used to measure the characteristic X-ray emission spectra. Wavelength dispersive X-ray fluorescence (WDXRF) analyses depend upon the use of diffracting crystal to determine the characteristic wavelength of the emitted X-rays. Energy dispersive X-ray fluorescence (EDXRF) employs detectors that directly measure the energy of the X-rays by collecting the ionization produced in suitable detecting medium.

X-ray emission spectra are known to be influenced by chemical combination of X-ray emitting atoms with different ligands. The effect of the chemical combination, however are not large and a theoretical interpretation of these effects has not been established completely. Therefore, chemical effects have rarely been utilized in the characterization of materials. The purpose of this work was to study chemical effects and discuss their applications to Yttrium (Y) in various compounds. So much so that, this paper presents and discusses the measured spectra both energy dispersive and wave-length dispersive X-ray spectrometer. In the first part of the study, the effect of the 0.6T and 1.2T external magnetic field and chemical state on the Ka, $K\beta_{1,3}$ and $K\beta_{2,4}$ X-ray energies and relative intensity ratios for Y, YBr_3, YCl_3, YF_3, $Y(NO_3)_3.6H_2O$, Y_2O_3, YPO_4, $Y(SO_4)_3.8H_2O$ and Y_2S_3 have been investigated, using the 22.69 keV X-rays from a ^{109}Cd and 59.54 keV γ-ray from a ^{241}Am as photon sources. The measurements were done using an energy dispersive Si(Li) detector with photon excitation by radioisotopes. For B=0, the present experimental results were compared with the experimental and theoretical data in the literature.

The results show that Y_2O_3, YF_3 and Y_2S_3 can change owing to the applied magnetic field. In addition, we found that the energy of characteristic X-ray series is totally independent of the excitation source and mode. However, changes have been observed in X-ray spectra when the element studied in the sample is chemically bonded to others. The development of high resolution spectrometers allows for the characterization and study of these effects.

In the second part of the study, energies and full width at half maximum (FWHM) values of the Ka, $K\beta_{1,3}$ and $K\beta_{2,4}$ X-ray of Y and its compounds were measured by a wavelength dispersive spectrometer. An accurate analytical representation of each line, obtained by a fit to a minimal set of Gaussians, is presented. The absolute energies and FWHM values derived from the data, agree well with previous measurements. Possible origins of chemical shifts are discussed. It was found that the chemical shifts of Y Ka line in pure Y and its some compounds relatively small (less than 0.1 eV with pure Y as reference). The influence of crystal symmetry on the energy shifts of X-ray lines is an interesting aspect of our study. The results demonstrate a clear dependence of the energy shifts on the chemical state of the element in the sample. The relative intensities are more susceptible to the chemical environment than the energy shifts.

It is well known that the chemical environment of an element affects and modifies the various characteristics of its X-ray emission spectrum. Most of the works suffer from neglecting chemical influences, and usually theoretical atomic values (Scofield, 1974a, 1974b) are used as a reference even for quite different chemical compounds of certain element. However, some papers deal with chemical effects (Berenyi et al., 1978; Rao et al., 1986), mostly in connection with X-ray emission after an electron capture process (EC) and partially after photoionisation (PI). Paic &Pecar (1976) found that for first-row transition elements the $K\beta/Ka$ ratio depends on the mode of excitation. The difference between the ratios for electron-capture decay and photoionization becomes almost 10%. Similar results were obtained by Arndt et al., (1982) and they pointed out that the difference comes from a strong shake-off process accompanying photoionization.

The $3d$ transition metals have played an important role in the development of modern technology, and knowledge of their valence electronic structure is very important for understanding their physical properties. X-ray spectroscopy is an established tool for probing the electronic structure of $3d$ transition metal compounds (Meisel et al., 1989). A number of techniques, such as photoemission spectroscopy, X-ray absorption and X-ray emission spectroscopy create a hole in an inner shell in order to investigate the valance

Determination of Chemical State and External Magnetic Field Effect on the Energy Shifts and X-Ray Intensity
Ratios of Yttrium and Its Compounds

87

electron configuration. Although some investigations have been made to study their electronic structures individually, no systematic study has been made so far for understanding the valence electronic structure of all the $3d$ transition metals. With a deeper theoretical understanding of the underlying processes and further improving X-ray sources, sophisticated experiments have been developed (e.g., resonant inelastic scattering, magnetic dichroism (Groot, 1994a,b)) that give detailed information on the valance electron configuration.

In a number of X-ray spectral studies of $3d$ transition metals it has been observed that the $K\beta$-to-$K\alpha$ X-ray intensity ratios are dependent on the physical and chemical environments of the elements in the sample. In the earlier studies of $3d$ metal compounds (Küçüköder et al., 1993; Padhi et al., 1993, 1995), the influence of chemical effects has shown difference in the $K\beta$-to-$K\alpha$ X-ray intensity ratios up to nearly 10%. Such chemical effects can be caused either by a varying $3d$ electron population or by the admixture of p states from the ligand atoms to the $3d$ states of the metal or both. Brunner et al. (1982) explained their experimental results by the change in screening of $3p$ electron by $3d$ valance electrons as well as the polarization effect. They also pointed out that the chemical effect is almost the same order of magnitude as the effect of excitation mode and both effects should be studied separately. However, most of these measurements have been performed with solid-state X-ray detectors and the change in the satellite peaks in the $K\beta$ X-ray region has not been studied because of poor energy resolution. Urch (1979) discussed the chemical effect on the K X-ray spectra based on molecular-orbital (MO) theory. Similar studies on the chemical effect on the X-ray spectra have already been done extensively. However, these studies are concerned mostly on with transition energies and profiles of X-rays, and qualitative discussions on the intensities have not yet been made. Tamaki et al. (1979) studied Cr and 55Mn-labeled compounds and reported that the $K\beta/K\alpha$ ratio increases with increasing formal oxidation number of the element in the compound. Kataria et al. (1986) found deviations of up to 10% for the same ratio in the case of Mn compounds. Mukoyama et al. (1986) experimentally confirmed the theoretical predictions following Brunners' (1982) model in the case of Te and Mo compounds for $K\beta_{1,3}$ and $K\beta_2$ components.

Wide employed applications and the intriguing asymmetry of the Cu Kα and Kβ line shapes (Deutsch&Hart, 1982) along with those of all 3d transition elements, led in turn to a century of extensive spectrometric studies of the Cu Kα and Kβ spectra. In spite of these extensive studied, recent studies reveal that surprises still lurk under the skewed $K\alpha_{1,2}$ and overlapping $K\beta_{1,3}$ lines, and the related multi-electronic satellite (S) and hypersatellite (HS) spectra. The asymmetric lineshape of the copper emission lines were attributed in the past to a number of different processes: Kondo-like interaction of the conduction electrons with the core holes, final state interactions between the core holes and the incomplete $3d$ shell, $2p/3d$ shell electrostatic exchange interaction, and most importantly, shake-up and shake-off of electrons from the $3l$ shells. The last process, in particular, received in the past strong experimental support.

Raj et al. (1998) were carried studies on CrB, CrB2 and FeB forms in order to look into the electronic structure of the transition metals in monoborides and diborides. In order to understand the valence electronic structure of the transition metals in the compounds, they have tried to compare the measured $K\beta$-to-$K\alpha$ ratios with the multiconfiguration Dirac±Fock calculations assuming different electronic configurations for the transition metal. Such a comparison would provide information on the valence electronic structure of the transition metals in the compounds, which could in turn provide information on the rearrangement of electrons between $3d$ and $4s$ states of the metal or electron transfer from the $3d$ state of the metal to the ligand atoms or vice-versa.

The chemical environment has a strong effect on the transitions originated in valence band and its influence could clearly be observed in the emission spectrum structure. The P-$K\beta$ spectrum has been studied by many authors (Takashi, 1972; Taniguchi 1984; Torres Delluigi et al., 2003), who used both single-crystal and two-crystal spectrometers with conventional X-ray sources. These authors showed some modifications in the $K\beta$ spectra and its relation with P chemical environment. Compounds with oxygen as ligand atom, a relationship between the ratio of the $K\beta'$ line intensity to the total intensity of the $K\beta$ line and the energy shift of the $K\alpha1,2$ lines was found by them. Fichter (1975) discussed the $K\alpha$-line shifts related to the oxidation number of the P-atom. The chemical shift of X-ray emission lines is usually interpreted with the effective charges or oxidation number of the X-ray emitting atom (Leonhardt&Meisel, 1970; Meisel et al., 1989). For example, the Al $K\alpha$ lines shift to higher energy in going from the metal to the oxide (Nagel et al., 1974). By comparing the measured chemical shifts with those of the reference compounds, Gohshi et al. (1973, 1975) determined the chemical state of S, Cr and Sn. They obtained not only qualitative, but also quantitative results.

Theoretical studies of emission spectra were performed mostly to study atoms with simple electronic configurations (see, e.g., the review by Mukoyama et. al., 2004). Theoretical calculations for solids and molecules have been done mainly to predict transition energies and line profiles, but evaluation of transition probabilities is rather scarce. This is due to two reasons: Firstly, molecular orbital methods and band theories are originally developed for ground states and sometimes difficult to apply to excited states with an inner-shell vacancy. Secondly, matrix elements for absorption and emission processes in molecules include multi-center integrations, which are tedious and require long computing times. Most individual authors indicate that their results favor the Dirac-Hartree-Fock calculations of Scofield (1974a), rather than the significantly lower predictions of the same author's earlier Dirac-Hartree-Slater calculations (Scofield, 1969). Both of these describe the de-excitation of a single K vacancy in a neutral atom. However careful examinations (Salem et al., 1974; Khan&Karimi, 1980) of all available data reveal a tendency for $K\beta/K\alpha$ to fall somewhat below the DHF predictions in the atomic number region $21<Z<32$ where the $3d$ subshell is filling.

Band et al. (1985) applied the scattered-wave (SW) $X\alpha$ MO method to calculate the chemical effect on the $K\beta/K\alpha$ intensity ratios. They performed the MO calculations for different chemical compounds of Mn and Cr using the cluster method and obtained the spherically averaged self-consistent potential and the total charge of the valance electrons in the central atom region. Chemical effect on the $K\beta/K\alpha$ X-ray intensity ratios or some Mn and Cr compounds has been studied both theoretically and experimentally by Mukoyama et al. (1986). The K X-ray spectra were measured by the use of a double crystal spectrometer with high energy resolution. The theoretical calculations were made with the use of the discrete-variational $X\alpha$ molecular-orbital method and the X-ray intensities were evaluated in the dipole approximation using molecular wave functions. Mukoyama et al. (2000) have calculated the electronic structures of tetraoxo complexes of $4d$ and $5d$ elements with the discrete-variational $X\alpha$ (DV-$X\alpha$) MO method. They found that the for Tc compounds, the calculated values were in good agreement with the measured values. In the case of Mo K X-rays, the agreement theory and experiment is not as good as with Tc compounds. Yamoto et al. (1986) studied the variation of the relative K X-ray intensity ratios for compounds involving Tc isotopes, [95m]Tc, [97m]Tc and [99m]Tc. They found that the chemical effect on the $K\beta/K\alpha$ ratios for $4d$ elements is small but the dependence of the $K\beta_2/K\alpha$ ratios on the chemical environments is appreciable.

Mukoyoma et al. (1986) have calculated the $K\beta_2/K\alpha$ intensity ratios for chemical compounds of $4d$ transition elements by the use of the simple theoretical method of Brunner et al. (1982), originally developed for $3d$ elements. Although they obtained good agreement between theories and experimental, it was found that their model is inadequate for the metallic cases.

These investigations on the effect of $3d$ and $4d$ electrons were performed only to understand the chemical effect on the X-ray intensity ratios. However, if the dependence on the excitation mode is also caused by the difference in the number of $3d$ electrons, as shown in our previous work, both effects, i.e. the dependence on the chemical environment and on the excitation mode, can be treated simultaneously to estimate the $K\beta/K\alpha$ ratios in terms of the number of $3d$ electrons. However it may also be possible that these ratios are also expressed as a function of other parameters, such as bond length and effective number of $4p$ electrons. Considering these facts, it is interesting to study the dependence of the $K\beta_2/K\alpha$ ratio in $3d$ elements on various parameters of chemical compounds.

Iiahara et al. (1993) measured the L X-ray intensity ratios for some Nb and Mo compounds. When the measured $L\gamma_1/L\beta_1$ ratios were plotted as a function of the effective number of $4d$ electrons, they found that the experimental data are experimental data are almost on a straight line. However, it should be noted that the $4d{\to}2p$ transitions are allowed dipole transition and the 4d electron is the valance shell electron which participates directly in the X-ray emission. In this case the X-ray emission rate is proportional to the number of $4d$ electrons and increases with increasing effective number of $4d$ electrons.

The chemical behavior of actinide atoms (in particular, that of uranium) is determined by valance nl-electrons of three types: 7s, 6d and 5f. Although the bond energies of these electrons are almost equal, their wave-function differs greatly in distribution in the radial direction (Katz et al., 1986; Balasubramanian et al., 1994). It can be said that the 5f electrons have an only core arrangement in the atom. Therefore, when actinides chemical bonding is studied, several questions should be raised: (1) the possibility and form of 5f electrons participation in chemical bonding; (2) the necessity for taking into account the splitting of valance levels of the atom into two sublevels nl+ and nl- with total angular momentum j=1±1/2 because of the relativistic effect of spin-orbital splitting (SOS) (Pyykko,1988; Pepper et al., 1991); (3) the energetic stabilization of the specific chemical state of the heavy atom due to fine effects of electron density redistribution on valance orbital; (4) the possibility of independent participation of split subshells in chemical bond formation. One of the methods of modern precise spectroscopy capable of providing a correct description of chemical bonding process is the chemical shift (CS) method of X-ray emission lines, i.e. the change in their energy when the chemical state of the emitting atom is changed (Gohshi&Ohtsuka, 1973; Makarov, 1999; Batrakov et al., 2004).

Atomic theory has shown that the magnetic dipole moments observed in bulk matter arise from one or two origins: one is the motion of the electrons about their atomic nucleus (orbital angular momentum) and the other is the rotation of the electron about its own axis (spin angular momentum). The nucleus itself has a magnetic moment. Except in special types of experiments, this moment is so small that it can be neglected in the consideration of the usual macroscopic magnetic properties of bulk matter. When the atom is placed in an external magnetic field, the magnetic field produces a torque on the magnetic dipole. The torque is tending to align the dipole with the field, associated with this torque; there is a potential energy of orientation:

$$\Delta E = -\mu_l B \tag{1}$$

μ_l is the orbital magnetic dipole moment of an electron. According to the quantum theory, all spectral lines arise from transitions of electrons between different allowed energy levels within the atom and the frequency of the spectral line is proportional to the energy difference between the initial and final levels. The slight difference in energy is associated with these different orientations in the magnetic field. In the presence of a magnetic field, the elementary magnetic dipoles, whether permanent or induced, will act to set up a field of induction of their own that will modify the original field.

Today investigations of magnetic effects on X-ray spectra became actual both from theoretical and experimental points of view. The numbers of works on this subject deal with magnetic circular dichroism (MCD) in X-ray absorption spectroscopy (XAS), that gives information on empty electron states in a valence band and their spin configurations (Thole et al., 1992, Stöhr&Wu, 1994). Several experiments have been performed on the external magnetic field effect on the K shell X-ray emission lines. Demir et al., (2006a) determined how the radiative transitions and the structures of the atoms in a strong magnetic field are affected, $K\alpha$ and $K\beta$ X-ray production cross sections, the K-shell fluorescence yields and $I(K\beta/K\alpha)$ intensity ratios for ferromagnetic Nd, Gd, and Dy and paramagnetic Eu and Ho were investigated using the 59.5 keV incident photon energy in the external magnetic fields intensities ± 0.75 T. On the other hand, Demir et al., (2006b) measured L_3 subshell fluorescence yields and level widths for Gd, Dy, Hg and Pb at 59.5 keV incident photon energy in the external magnetic field of intensities ± 0.75 T. Porikli et al. (2008a; 2008b; 2008c) conduct measurements using pure Ni, Co, Cu and Zn and their compounds. Characteristic quantities such as position of line maxima, full widths at half maximum (FWHM), indices of asymmetry and intensity ratio values were determined in the values of external magnetic field 0.6 T and 1.2 T. Several experiments have been performed on the external magnetic field effect on the K shell X-ray emission lines. Commonly, experimental L X-ray intensities are measured using radioisotopes as excitation sources (Han et al., 2010; Porikli, 2011b). They have the advantages of stable intensity and energy and of small sizes, which allow compact and efficient geometry, and they operate without any external power.

Our motivation in performing this experiment has been two fold. First, with the aim of a better understanding of the chemical effect and external magnetic field effect, we conduct measurements using pure yttrium (Y) and its compounds. Characteristic quantities such as position of line maxima, full widths at half maximum (FWHM), indices of asymmetry and $K\beta_1/K\alpha$, $K\beta_2/K\alpha$, $K\beta_2/K\beta_1$ and $K\beta/K\alpha$ intensity ratio values are determined in the values of external magnetic field 0.6 T and 1.2 T. In the present work, the measurements were done using a filtered 22.69 keV from Cd-109 and 59.54 keV from Am-241 point source and Si(Li) detector. Particle size effects were circumvented. Peak areas were determined using Gaussian fitting procedures and the errors in various corrections such as self-absorption and detector efficiency were minimized. The measured values were compared due to the external magnetic field and chemical effect. The measured values for B=0 were compared with other experimental and theoretical results. To our knowledge, these intensity ratio values of Y in the external magnetic field have not been reported in the literature and appear to have been measured here for the first time. Secondly, spectra of K X-rays emitted from a Y target were measured in high resolution wave-length dispersive X-ray spectrometer (WDXRF). After the measurement, characteristic quantitative such as peak energy, indices of asymmetry, FWHM are determined. The measured spectra were described in terms of a background function (a straight line) and peaks having Gaussian profiles. The Microcal Orgin 7.5 was used for peak resolving and background subtraction of K X-rays.

2. Experimental

2.1 Experimental set up (EDXRF)

Yttrium compounds can serve as host lattices for doping with different lanthanide cations and they used as a catalyst for ethylene polymerization. As a metal, it is used on the electrodes of some high-performance spark plugs. Yttrium is also used in the manufacturing of gas mantles for propane lanterns as a replacement for thorium, which is radioactive. Developing uses include yttrium-stabilized zirconia in particular as a solid electrolyte and as an oxygen sensor in automobile exhaust systems. Yttrium is used in the production of a large variety of synthetic garnets. Small amounts of yttrium (0.1 to 0.2%) have been used to reduce the grain sizes of chromium, molybdenum, titanium, and zirconium. It is also used to increase the strength of aluminium and magnesium alloys. The addition of yttrium to alloys generally improves workability, adds resistance to high-temperature recrystallization and significantly enhances resistance to high-temperature oxidation (see graphite nodule discussion below).

The studied elements were Y, YBr_3, YCl_3, YF_3, $Y(NO_3)_3.6H_2O$, Y_2O_3, YPO_4, $Y(SO_4)_3.8H_2O$ and Y_2S_3. The purity of commercially obtained materials was better than 99%. For powdered samples, particle size effects have a strong influence on the quantitative analysis of infinitely thick specimens. Even for specimens of intermediate thickness, in which category the specimens analyzed in the present study fall, these effects can be significant. Therefore, to circumvent particle size effects all samples were grounded and sieved through a -400 mesh (<37 μm) sieve. The powder was palletized to a uniform thickness of 0.05-0.15 g cm^{-2} range by a hydraulic press using 10 ton in^{-2} pressure. The diameter of the pellet was 13 mm.

All of the lines were excited using a 100 mCi Am-241 annular radioactive source and Cd-109 point source of 10 mCi strength (providing 5.0×10^3 steradian^{-1} photon flux of Ag X-radiation). The fluorescent X-rays emitted from the targets were analyzed by a Si(Li) detector (effective area 12.5 mm^2, thickness 3 mm, Be window thickness 0.025 mm).

For each sample three separate measurements have been made just to see the consistency of the results obtained from different measurements agreed with a deviation of less than 1%. The experimental setup consist of a Si(Li) detector and Cd-109 radioactive source as shown in Fig. 1. The mechanical arrangement to house the source-sample-detector combination in a definite geometry was shown in Fig. 1. An Al, Pb conical collimator was used between the sample and the detector for the excitation to obtain a large beam of emergent radiation and to avoid the interaction of the X-rays emitted by the component elements of the radioactive capsule and detector.An Al, Pb conical collimator was used between the sample and the detector for the excitation to obtain a large beam of emergent radiation and to avoid the interaction of the X-rays emitted by the component elements of the radioactive capsule and detector. This collimator has an external diameter of 13 mm and it was placed in the internal diameter of the radioactive source (8 mm). A graded filter of Pb, Fe and Al to obtain a thin beam of photons scattered from the sample and to absorb undesirable radiation shielded the detector. The sample-detector and excitation source-sample distances were optimized to get maximum count rate in the fluorescent peaks. The sample was placed approximately at 45° to the source-plane as well as to the detector-plane so that the intensity of scattered radiation could be minimized (Giauque et al., 1973). The count rate kept below 1000 counts s^{-1} in order to avoid peak broadening, energy shift and non-linearity. The data were collected into 16384 channels of a digital spectrum analyzer DSA-1000. The energy per channel was adjusted as 4 eV to determine the peak centroids and to discriminate the overlapped peaks. The samples were mounted in a sample holder placed between the pole pieces of an electromagnet capable of producing the magnetic field of approximately 2.66T at 2 mm pole range. During the study, the magnetic field intensities of, 0.6 T and 1.2 T were applied to the samples. An ammeter monitored the continuity and stability of the currents feeding the electromagnet. A typical K X-ray spectrum of Y at the 0.0 T, 0.6 T and 1.2 T is shown in Fig. 2. A typical $K\alpha$, $K\beta_{1,3}$ and $K\beta_{2,4}$ spectrum of Y, YBr$_3$, YCl$_3$, Y(SO$_4$)$_3$.8H$_2$O and Y$_2$S$_3$ are shown in Fig. 3.

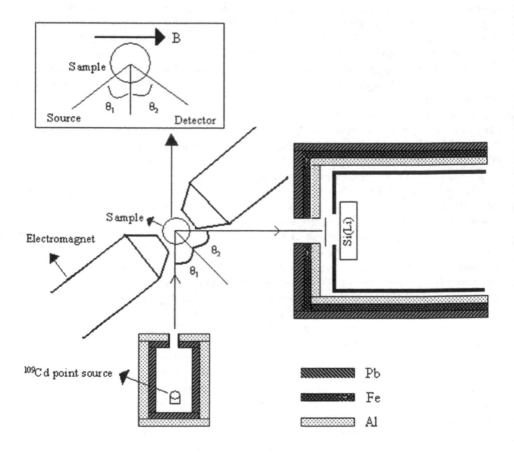

Fig. 1. Experimental set-up.

Determination of Chemical State and External Magnetic Field Effect on the Energy Shifts and X-Ray Intensity
Ratios of Yttrium and Its Compounds

93

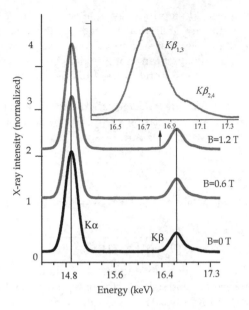

Fig. 2. A typical K X-ray spectrum of the Y target in B=0, B=0.6T and B=1.2 T magnetic field.
The spectra were plotted after smoothing.

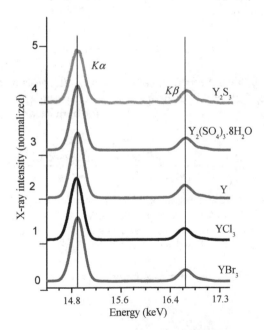

Fig. 3. Measured Ka, $K\beta_{1,3}$ and $K\beta_{2,4}$ spectra of Y, YBr₃, YCl₃, Y(SO₄)₃.8H₂O and Y₂S₃. The
spectra were plotted after smoothing.

Spectrum evaluation is a crucial step in X-ray analysis, as much as sample preparation and quantification. As with any analytical procedure, the final performance of X-ray analysis is determined by the weakest step in the process. The processing of ED spectra by means of computers has always been more evident because of their inherent digital nature. Due to relatively low resolving power of the employed Si(Li) detector, the process of evaluating XRF spectra is prone to many errors and requires dedicated software. For this purpose a software package called ORGIN was used for peak resolving background subtraction and determination of the net peak areas of K X-rays which is based on the non-linear least squares fitting of a mathematical model of the XRF spectrum.

2.2 Data analysis (EDXRF)

The $K\beta/K\alpha$ X-ray intensity ratio values have been calculated by using the relation

$$\frac{I(K\beta)}{I(K\alpha)} = \frac{N(K\beta)}{N(K\alpha)} \frac{\varepsilon(K\alpha)}{\varepsilon(K\beta)} \frac{\beta(K\alpha)}{\beta(K\beta)} \tag{2}$$

where $N(K\alpha)$ and $N(K\beta)$ are the net counts under the $K\alpha$ and $K\beta$ peaks, respectively. $\beta(K\alpha)$ and $\beta(K\beta)$ are the self-absorption correction factor of the target and $\varepsilon(K\alpha)$ and $\varepsilon(K\beta)$ are the detector efficiency for $K\alpha$ and $K\beta$ rays. The values of the factors, $I_0 G\varepsilon$ which contain terms related to the incident photon flux, geometrical factor and the efficiency of the X-ray detector, were determined by collecting the $K\alpha$ and $K\beta$ X-ray spectra of Ti, As, Br, Sr, Y, Zr and Ru with the mass thickness 0.02-0.17 g/cm^2 in the same geometry and calculated by using the following equation

$$I_0 G\varepsilon_{Ki} = \frac{N_{Ki}}{\sigma_{Ki}\beta_{Ki}t_i} \tag{3}$$

where N_{Ki} and β_{Ki} ($i=\alpha, \beta$) have the same meaning as in Eq. (2). σ_{Ki} is X-ray fluorescence cross-section, G is a geometry factor and t is the mass of the sample in g/cm^2.
The self absorption correction factor β is calculated for both $K\alpha$ and $K\beta$ separately by using the following expression

$$\beta_{Ki} = \frac{1 - \exp\{-[\mu(E_0)\sec\theta_1 + \mu_{Ki}(E)\sec\theta_2]t\}}{[\mu(E_0)\sec\theta_1 + \mu_{Ki}(E)\sec\theta_2]t} \tag{4}$$

where $\mu(E_0)$ and $\mu_{Ki}(E)$ are the total mass absorption coefficients taken from WinXCOM programme which is the Windows version of XCOM. XCOM is the electronic version of Berger and Hubbell's Tables (Berger et al., 1987). The angles of incident photons and emitted X-rays with respect to the normal at the surface of the sample θ_1 and θ_2 were equal to 45° in the present setup.
The term σ_{Ki} represents the K X-ray fluorescence cross-sections and is given by

$$\sigma_{Ki} = \sigma_K^P w_K f_{Ki} \tag{5}$$

σ_K^P is the K shell photo ionization cross-section (Scofield, 1973), w_K is the fluorescence yield (Krause et al., 1979) and f_{Ki} is fractional X-ray emission rate (Scofield, 1974a).

Determination of Chemical State and External Magnetic Field Effect on the Energy Shifts and X-Ray Intensity
Ratios of Yttrium and Its Compounds

95

2.3 Experimental set up (WDXRF)

A commercial WDXRF spectrometer (Rigaku ZSX 100e) was used for analysis of the different samples. This instrument is usually equipped with a 3 kW Rh-anode tube working at a voltage range of 20–50 kV and a current from 20 to 50 mA. It is possible to use primary beam filters (made of Zr, Al, Ti or Cu) between the primary radiation and the sample holder to reduce the background continuum and to improve the signal-to-noise ratio. Energy resolution and efficiency for each analytical line also depend on the collimator aperture and the analyzer crystal in use. Several different collimators can be used to reduce the step/scan resolution, as well as up to ten analyzer crystals, to better enhance spectral data for a specific element. Detection can be performed using a flow proportional counter (light elements) or a scintillation counter (heavy elements). In this work, analyses were made in vacuum atmosphere. Moreover, to avoid possible problems with inhomogeneity when measuring the samples, a sample spinner facility was used in all cases.

To investigate the spectrometer sensitivity in measuring of intensity and energy shift, one sample at same conditions was measured for three times. Because of the use of instruments such as sieve weight and hydraulic press, errors are caused in the results of analysis. These errors were called manual and instrumental errors. Three samples were prepared and measured for same conditions to determine these errors.

2.4 Spectral profile analysis (WDXRF)

The common method for evaluation of spectra in WDXRF is by the use of net peak line intensity. This is due to the high efficiency in the analytical results from the scintillation and/or the flow counter detectors. These detectors can receive up to 2×10^6 cps. In contrast, the common spectra evaluation in EDXRF is based on integration of the gross or net peak area due to a lower efficiency in the solid state detectors, usually limited to a maximum count of 5×10^4 cps. Taking into account these facts and to improve the sensitivity of the signal, the spectral data obtained by the WDXRF equipment were treated using the deconvolution software (Microcal Orgin 7.5), traditionally used in EDXRF spectrometry, to obtain the peak areas. The total number of counts increases considering the total peak area instead of only the analytical line. This leads also to an improvement of sensibility and detection limits. Once samples were analyzed, the identification of elements from the WDXRF spectra was done by using the qualitative scanning mode linked to the equipment, which includes automatic peak and element identification. The principle of WDXRF spectrometry is the use of different analyzer crystals to diffract and separate the different characteristic wavelengths of the elements present in the sample. For that reason, in WDXRF measurements, a multi-spectrum was obtained resulting from the use of different analyzing crystals, excitation conditions, etc.

Rigaku has improved their semi-quantitative software package further with the introduction of SQX. It is capable of automatically correcting for all matrix effects, including line overlaps. SQX can also correct for secondary excitation effect by photoelectrons (light and ultra-light elements), varying atmospheres, impurities and different sample sizes.

The obtained multispectra were split into the different individual spectra and were converted to energies by inversion of the channels to be treated using the means of the SQX software to perform spectral deconvolution and fitting and to evaluate element net peak areas from the spectra. Peak fitting was done by iteration to better adjust the peak and the background to minimize the chi-square of the fitting on each spectra. Fig. 4 shows the spectrum of Y. Measured numbers of counts are shown as solid black circles, while the red line represents the overall fit. The background is shown as a blue line.

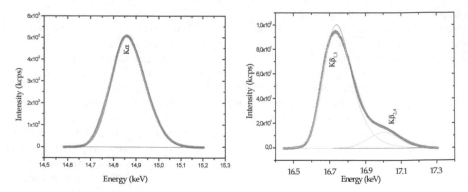

Fig. 4. Solid circles: Measured spectrum of Y K X-rays. Lines: Overall fitting function (red) and its components (green).

3. Results and discussion

A frequently used and convenient (but not very accurate) quantification of the line shape is by its full width at half maximum (FWHM) and index of asymmetry. These characteristics of lines (line parameters) are sensitive to change: line position (chemical shift), line shape (full width at half maximum (FWHM) and index of asymmetry) and additionally mutual ratios of line intensities. The peak position was determined at the center point of the 9/10 intensity of the smoothed line shape as illustrated in Fig. 5. It was known from our experience that the standard deviation of the peak position was determined using the peak top. Parameters such as FWHM and asymmetry index, defined in Fig. 5, were evaluated using the smoothed data. The Savitzky-Golay smoothing method was iteratively processed one time. Spectral smoothing was important for reducing the standard deviation of these parameters.

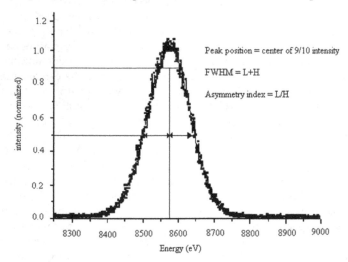

Fig. 5. Definition of asymmetry index, FWHM and peak position determined from 9/10 intensity.

The FWHM values for all compounds investigated for Cd-109 and Am-241 radioactive sources are given in Tables 1 and 2. Among all the Y compounds, the $Y(NO_3)_3.6H_2O$ Ka emission line shows large widths for both lines. Both the Ka, $K\beta_{1,3}$ and $K\beta_{2,4}$ lines become narrow when the compounds crystal system is monoclinic. In all cubic compounds such as Y_2O_3 and Y_2S_3 the lines FWHM values come closer to pure Y FWHM value. As can be seen from Table 1 and 2, for Zn compounds, the variation in the values of FWHM is relatively small. But when we compare the Y compounds with pure forms, we realize changes in both Ka, $K\beta_{1,3}$ and $K\beta_{2,4}$ FWHM values.

The nonmonotonic behavior of the Ka widths is probably due to the behavior of the L levels widths rather than the K level ones. The smaller overlap of the M and K wave functions, as compared to the K and L ones, may reduce the relative influence of possible similar sized nonmonotonic contributions originating in the final state level widths. Since experimental results FWHM and the index of asymmetry for $B\neq0$ cannot be found in the literature, the comparison is not made with the other experimental values. As can be seen from Table 1 and 2, all FWHM values systematically decrease with increasing magnetic field intensity. Table 1 and 2 show that the experimental values of FWHM are unity within experimental uncertainties, suggesting the absence of chemical and external magnetic field effects.

Element	External Magnetic Field	Asymmetry Index (eV)			FWHM (eV)	
		Ka	$K\beta_{1,3}$	$K\beta_{2,4}$	Ka	$K\beta_{1,3}$
Y	B=0	1.125	1.136	1.008	3.232	4.333
	B=0.6T	1.079	1.112	1.006	3.223	4.268
	B=1.2T	0.991	0.109	0.996	3.057	4.016
$Y(NO_3)_3.6H_2O$	B=0	1.122	1.127	1.011	3.211	4.228
	B=0.6T	0.997	1.103	1.009	3.170	4.220
	B=1.2T	0.968	1.002	0.994	3.109	4.103
YCl_3	B=0	1.110	1.119	0.994	3.197	4.267
	B=0.6T	1.017	1.077	0.991	3.133	4.209
	B=1.2T	0.984	1.004	0.985	3.110	4.111
YPO_4	B=0	1.114	1.111	0.987	3.199	4.337
	B=0.6T	0.983	1.071	0.967	3.139	4.259
	B=1.2T	0.980	1.002	0.956	3.062	4.058
YBr_3	B=0	1.098	1.078	0.971	3.201	4.284
	B=0.6T	1.007	1.006	0.970	3.187	4.065
	B=1.2T	1.000	0.989	0.944	3.110	3.943
Y_2O_3	B=0	1.071	1.05	0.938	3.266	4.121
	B=0.6T	1.010	1.001	0.930	3.133	4.043
	B=1.2T	0.994	0.983	0.915	3.109	3.997
YF_3	B=0	1.099	1.077	0.933	3.245	4.264
	B=0.6T	1.022	1.011	0.929	3.120	4.001
	B=1.2T	0.988	0.984	0.886	3.100	3.966
$Y(SO_4)_3.8H_2O$	B=0	1.055	1.035	0.921	3.189	4.166
	B=0.6T	1.003	1.004	0.917	3.166	3.976
	B=1.2T	0.969	0.966	0.910	3.037	3.874
Y_2S_3	B=0	1.024	1.031	0.910	3.337	4.494
	B=0.6T	1.012	0.994	0.883	3.266	4.441
	B=1.2T	0.985	0.981	0.867	3.190	4.284

Table 1. Full width at half maximum (FWHM) and asymmetry index values of Ka, $K\beta_{1,3}$ and $K\beta_{2,4}$ emission lines in Y compounds for Cd-109 radioactive source.

To obtain more definite conclusions on FWHM dependency of the external magnetic field, more experimental data are clearly needed. The experimental uncertainties are always <0.05 eV for the FWHM.

According to Allinson (1933), the index of asymmetry of an X-ray emission line is defined as the ratio of the part of the FWHM lying to the long-wavelength side of the maximum ordinate to that on the short-wavelength side. In Table 1 and 2, the index of asymmetry for Ka, $K\beta_{1,3}$ and $K\beta_{2,4}$ emission lines are presented. The experimental uncertainties in the values cited in the table were determined taking into account multiple measurements and multiple fits of each spectrum. The errors for the index of asymmetry are ≤ 0.1 eV for Ka, $K\beta_{1,3}$ and $K\beta_{2,4}$.

Element	External Magnetic Field	Asymmetry Index (eV)			FWHM (eV)	
		Ka	$K\beta_{1,3}$	$K\beta_{2,4}$	Ka	$K\beta_{1,3}$
Y	B=0	1.120	1.113	1.022	3.229	4.297
	B=0.6T	1.089	1.100	1.016	3.212	4.203
	B=1.2T	1.009	0. 989	0.984	3.050	4.100
Y(NO$_3$)$_3$.6H$_2$O	B=0	1.113	1.110	1.020	3.114	4.116
	B=0.6T	1.028	1.104	1.013	3.017	4.111
	B=1.2T	0.996	0.993	1.001	3.091	4.075
YCl$_3$	B=0	1.109	1.105	1.017	3.077	4.122
	B=0.6T	1.003	1.007	1.009	3.041	4.110
	B=1.2T	0.978	0.987	0.987	3.000	4.004
YPO$_4$	B=0	1.111	1.075	0.993	3.201	4.177
	B=0.6T	0.989	1.032	0.984	3.126	4.170
	B=1.2T	0.969	1.008	0.953	3.013	4.005
YBr$_3$	B=0	1.101	1.031	0.988	3.421	4.136
	B=0.6T	1.017	1.004	0.980	3.234	4.065
	B=1.2T	0.994	0.971	0.965	3.048	3.993
Y$_2$O$_3$	B=0	1.056	1.006	0.974	3.555	4.008
	B=0.6T	1.031	0.999	0.972	3.229	3.989
	B=1.2T	0.987	0.977	0.954	3.096	3.974
YF$_3$	B=0	1.077	1.001	0.970	3.301	4.123
	B=0.6T	1.005	0.984	0.954	3.137	4.012
	B=1.2T	0.993	0.975	0.950	3.009	3.940
Y(SO$_4$)$_3$.8H$_2$O	B=0	1.064	0.991	0.964	3.202	4.109
	B=0.6T	1.001	0.990	0.966	3.113	3.989
	B=1.2T	0.978	0.946	0.956	3.074	3.866
Y$_2$S$_3$	B=0	1.011	0.995	0.949	3.441	4.301
	B=0.6T	1.010	0.994	0.937	3.368	4.039
	B=1.2T	0.974	0.961	0.921	3.222	3.974

Table 2. Full width at half maximum (FWHM) and asymmetry index values of Ka, $K\beta_{1,3}$ and $K\beta_{2,4}$ emission lines in Y compounds for Am-241 radioactive source.

When the crystal system of Y compounds is cubic, the Ka emission lines are almost symmetric for Y_2O_3 and Y_2S_3. It is clear from results that, except for Y and $Y(NO_3)_3.6H_2O$,the asymmetry are generally larger for the Ka peak. So we can say that, the line shapes of Ka are not symmetric. It is also found from the Table 2, when the crystal system of Y compounds is trigonal (YBr$_3$), the Ka, $K\beta_{1,3}$ and $K\beta_{2,4}$ emission lines are almost symmetric too. For Am-241, line shapes are more symmetric than the Cd-109. As seen from Table 1 and 2, in the presence of an external magnetic field, the asymmetry index of the Y compounds change. The Ka, $K\beta_{1,3}$ and $K\beta_{2,4}$ emission lines asymmetry indices values decrease with external magnetic field. However, a more asymmetric structure is encountered for the elements of which their crystal symmetry is cubic. Also, it is observed that monoclinic group is more symmetric than the others.

Element	External Magnetic Field	Chemical shift (ΔE) (eV)			Energy shift (δE) (eV)		
		Ka	$K\beta_{1,3}$	$K\beta_{2,4}$	Ka	$K\beta_{1,3}$	$K\beta_{2,4}$
Y	B=0	0	0	0	0	0	0
	B=0.6T				0.074	0.109	0.101
	B=1.2T				0.161	0.260	0.237
Y(NO$_3$)$_3$.6H$_2$O	B=0	-0.333	-0.441	-0.216	0	0	0
	B=0.6T				0.086	0.098	0.115
	B=1.2T				0.141	0.137	0.227
YCl$_3$	B=0	-0.238	-0.309	-0.115	0	0	0
	B=0.6T				0.481	0.235	0.288
	B=1.2T				0.612	0.368	0.339
YPO$_4$	B=0	-0.121	-0.235	-0.103	0	0	0
	B=0.6T				0.335	0.279	0.336
	B=1.2T				0.455	0.355	0.399
YBr$_3$	B=0	-0.033	-0.065	-0.065	0	0	0
	B=0.6T				0.444	0.131	0.338
	B=1.2T				0.657	0.4	0.551
Y$_2$O$_3$	B=0	0.135	0.017	0.056	0	0	0
	B=0.6T				0.111	0.124	0.543
	B=1.2T				0.185	0.303	0.441
YF$_3$	B=0	0.164	0.191	0.198	0	0	0
	B=0.6T				0.167	0.166	0.112
	B=1.2T				0.533	0.529	0.144
Y(SO$_4$)$_3$.8H$_2$O	B=0	0.609	0.387	0.33	0	0	0
	B=0.6T				0.089	0.12	0.051
	B=1.2T				0.354	0.293	0.237
Y$_2$S$_3$	B=0	0.899	0.872	0.808	0	0	0
	B=0.6T				0.173	0.204	0.206
	B=1.2T				0.199	0.307	0.333

Table 3. Chemical shift (ΔE) and energy shift (δE) values of Ka, $K\beta_{1,3}$ and $K\beta_{2,4}$ emission lines in Y compounds for Cd-109 radioactive source.

The more unpaired 3d or 4d electrons the atom possesses, the more asymmetric will be the line observed. This kind observation led Tsutsumi (1959) to consider that the interaction between the hole created in the $2p_{3/2}$ or $2p_{1/2}$ shell (due to the transition of an electron from this shell to the 1s level) and the electrons in the incomplete 3d shell in the transition metal atoms is responsible for asymmetric nature of Ka lines. They proposed a theoretical model based on this idea to account for the asymmetry in the X-ray emission lines in the firs-row transition metal compounds. However, this is not the only consideration which can explain the origin of the asymmetry of the line; there are other considerations which are based on the relaxation effect of the inner state proposed by Parratt (1959) or on the interactions between 2p hole and electrons in the Fermi sea as proposed by Doniach and Sunjic (1970).

Element	External Magnetic Field	Chemical shift (ΔE) (eV)			Energy shift (δE) (eV)		
		Ka	$K\beta_{1,3}$	$K\beta_{2,4}$	Ka	$K\beta_{1,3}$	$K\beta_{2,4}$
Y	B=0	0	0	0	0	0	0
	B=0.6T				0.125	0.121	0.132
	B=1.2T				0.179	0.219	0.269
Y(NO$_3$)$_3$.6H$_2$O	B=0	-0.401	-0.347	-0.316	0	0	0
	B=0.6T				0.091	0.111	0.158
	B=1.2T				0.104	0.169	0.201
YCl$_3$	B=0	-0.023	-0.109	-0.153	0	0	0
	B=0.6T				0.226	0.254	0.259
	B=1.2T				0.559	0.472	0.318
YPO$_4$	B=0	0.231	0.304	0.221	0	0	0
	B=0.6T				0.553	0.602	0.436
	B=1.2T				0.598	0.556	0.511
YBr$_3$	B=0	-0.133	-0.227	-0.194	0	0	0
	B=0.6T				0.342	0.313	0.387
	B=1.2T				0.600	0.499	0.458
Y$_2$O$_3$	B=0	0.440	0.316	0.241	0	0	0
	B=0.6T				0.447	0.423	0.505
	B=1.2T				0.682	0.613	0.553
YF$_3$	B=0	0.206	0.391	0.190	0	0	0
	B=0.6T				0.367	0.276	0.211
	B=1.2T				0.533	0.590	0.438
Y(SO$_4$)$_3$.8H$_2$O	B=0	0.194	0.276	0.133	0	0	0
	B=0.6T				0.194	0.146	0.115
	B=1.2T				0.402	0.281	0.296
Y$_2$S$_3$	B=0	0.575	0.503	0.421	0	0	0
	B=0.6T				0.197	0.340	0.247
	B=1.2T				0.331	0.349	0.366

Table 4. Chemical shift (ΔE) and energy shift (δE) values of Ka, $K\beta_{1,3}$ and $K\beta_{2,4}$ emission lines in Y compounds for Am-241 radioactive source.

The chemical shift was the difference between the center point of the 9/10 peak intensity of a compound and that of pure Y measured before and after the measurement of the compound. When the environment of the emitting atom is changed, there are changes in the position of emission lines with respect to those in the pure metal. These changes are called chemical shifts. They are presented in Tables 3 and 4. Both the $4d$ electron configuration and crystal structure affect the chemical shift and energy shift with applied external magnetic field. There is a clear relationship with the external magnetic field values and the energy shift values of Y compounds, as is found in Table 3 and 4's last column. For higher values of external magnetic field, the values of energy shift increases systematically. But we do not find any relationship between external magnetic field and crystal structure of the compound. To obtain more definite conclusion, more experimental data for $4d$ compounds which crystal structure different are needed.

The errors of the chemical and energy shifts, originate mainly from the limited precision of our measurements, determined by repetitive measurements of all pure targets, which were prepared and placed in the same experimental geometry. The precision in the position of Ka, $K\beta_{1,3}$ and $K\beta_{2,4}$ lines was determined as 0.05 eV, whereas the maximum deviation of a single measurement from the average value was 0.1 eV.

Element	Differences between FWHM values [ΔFWHM=FWHM$_{com.}$-FWHM$_{pure}$] (eV)			Chemical shift (ΔE) (eV)		
	Ka	$K\beta_{1,3}$	$K\beta_{2,4}$	Ka	$K\beta_{1,3}$	$K\beta_{2,4}$
Y	0	0	0	0	0	0
Y(NO$_3$)$_3$.6H$_2$O	0.155	0.031	0.012	0.132	0.126	0.125
YCl$_3$	0.091	0.058	0.033	0.268	0.238	0.284
YPO$_4$	-0.223	-0.087	-0.107	-0.454	-0,444	-0.415
YBr$_3$	0.114	0.071	0.067	0.144	0.179	0.156
Y$_2$O$_3$	-0.077	-0.068	-0.021	-0.096	-0.126	-0.085
YF$_3$	-0.095	-0.034	-0.048	-0.115	-0.127	-0.167
Y(SO$_4$)$_3$.8H$_2$O	0.054	0.023	0.181	0.183	0.145	0.106
Y$_2$S$_3$	-0.013	-0.064	-0.079	-0.301	-0.397	-0.385

Table 5. Chemical shift (ΔE) and differences between FWHM values of Ka, $K\beta_{1,3}$ and $K\beta_{2,4}$ emission lines in Y compounds obtained for WDXRF.

It is also seen from Table 5 that the Ka line width of YPO$_4$ compound is wider than that of the other compounds. Compare with the Ka peak of the pure Y, that of cubic crystal structure Yttrium compounds shifted to lower energy, and the peak shift ordering was YF$_3$<Y$_2$O$_3$<Y$_2$S$_3$<YPO$_4$. The line shapes of Ka are generally symmetric. Y$_2$S$_3$ and Y$_2$O$_3$, where the $K\beta_{1,3}$ and $K\beta_{2,4}$ peak shifts are large, show prominent asymmetry.

The accurate knowledge of the $K\beta_1/Ka$, $K\beta_2/Ka$, $K\beta_2/K\beta_1$ and $K\beta/Ka$ intensity ratios is required for a number of practical applications of X-rays, e.g. molecular and radiation physics investigations, in non-destructive testing, elemental analysis, medical research etc. Therefore, these ratios depend sensitively on the atomic structure. Thus they have been widely used also for critical evaluation of atomic structure model calculations. We now discuss the values of these ratios as obtained in our measurements.

The relevant information in a spectrum is contained in its peaks whose position and area are linked respectively to the photon energy and the activity of the connected radionuclide. The peak areas can also be used to determine emission probabilities. In this work, peak areas were determined after the Ka, $K\beta_{1,3}$ and $K\beta_{2,4}$ areas were separated by fitting the measured spectra with multi-Gaussian function plus polynomial backgrounds using Microcal Orgin 7.5 software program. Details of the experimental set up and data analysis have been reported earlier (Porikli et al., 2011b).

Table 6 lists the theoretical values which were calculated by Scofield (Scofield 1974a; Scofield, 1974b). Addition to this, the measured values of the $K\beta_1/Ka$, $K\beta_2/Ka$, $K\beta_2/K\beta_1$ and $K\beta/Ka$ intensity ratios in Y, and previous experimental and the other theoretical values of these ratios for pure elements and their compounds are listed in Table 6.

Element	External Magnetic Field	Intensity Ratio	This Work	Scofield (1974a)	Manson &Kennedy (1974)	Ertuğral et al. (2007)
Y	B=0	$K\beta_{1,3}/Ka$	0.2307±0.010	0.22910		
		$K\beta_{2,4}/Ka$	0.0317±0.008	0.02902		
		$K\beta_{2,4}/K\beta_{1,3}$	0.1981±0.011	0.19220		
		$K\beta/Ka$	**0.1822±0.008**	**0.16960**	**0.1685**	**0.1856±0.009**
	B=0.6T	$K\beta_{1,3}/Ka$	0.2289±0.007			
		$K\beta_{2,4}/Ka$	0.0311±0.008			
		$K\beta_{2,4}/K\beta_{1,3}$	0.1956±0.011			
		$K\beta/Ka$	**0.1753±0.011**			
	B=1.2T	$K\beta_{1,3}/Ka$	0.2275±0.007			
		$K\beta_{2,4}/Ka$	0.0304±0.008			
		$K\beta_{2,4}/K\beta_{1,3}$	0.1941±0.011			
		$K\beta/Ka$	**0.1712±0.011**			
Y(NO₃)₃.6 H₂O	B=0	$K\beta_{1,3}/Ka$	0.2339±0.008			
		$K\beta_{2,4}/Ka$	0.0325±0.010			
		$K\beta_{2,4}/K\beta_{1,3}$	0.1987±0.011			
		$K\beta/Ka$	**0.1829±0.006**			
	B=0.6T	$K\beta_{1,3}/Ka$	0.2320±0.008			
		$K\beta_{2,4}/Ka$	0.0316±0.008			
		$K\beta_{2,4}/K\beta_{1,3}$	0.1977±0.010			
		$K\beta/Ka$	**0.1796±0.011**			
	B=1.2T	$K\beta_{1,3}/Ka$	0.2315±0.008			
		$K\beta_{2,4}/Ka$	0.0310±0.011			
		$K\beta_{2,4}/K\beta_{1,3}$	0.1954±0.011			
		$K\beta/Ka$	**0.1742±0.010**			
YCl₃	B=0	$K\beta_{1,3}/Ka$	0.2341±0.010			
		$K\beta_{2,4}/Ka$	0.0329±0.008			
		$K\beta_{2,4}/K\beta_{1,3}$	0.1992±0.009			
		$K\beta/Ka$	**0.1836±0.010**			
	B=0.6T	$K\beta_{1,3}/Ka$	0.2337±0.006			
		$K\beta_{2,4}/Ka$	0.0305±0.009			
		$K\beta_{2,4}/K\beta_{1,3}$	0.1952±0.009			

Determination of Chemical State and External Magnetic Field Effect on the Energy Shifts and X-Ray Intensity
Ratios of Yttrium and Its Compounds

103

Element	External Magnetic Field	Intensity Ratio	This Work	Scofield (1974a)	Manson &Kennedy (1974)	Ertuğral et al. (2007)
		$K\beta/K\alpha$	**0.1821±0.011**			
	B=1.2T	$K\beta_{1,3}/K\alpha$	0.2322±0.010			
		$K\beta_{2,4}/K\alpha$	0.0324±0.008			
		$K\beta_{2,4}/K\beta_{1,3}$	0.1972±0.009			
		$K\beta/K\alpha$	**0.1818±0.010**			
YPO₄	B=0	$K\beta_{1,3}/K\alpha$	0.2355±0.010			
		$K\beta_{2,4}/K\alpha$	0.0333±0.010			
		$K\beta_{2,4}/K\beta_{1,3}$	0.1999±0.010			
		$K\beta/K\alpha$	**0.1840±0.009**			
	B=0.6T	$K\beta_{1,3}/K\alpha$	0.2336±0.007			
		$K\beta_{2,4}/K\alpha$	0.0320±0.009			
		$K\beta_{2,4}/K\beta_{1,3}$	0.1966±0.011			
		$K\beta/K\alpha$	**0.1773±0.008**			
	B=1.2T	$K\beta_{1,3}/K\alpha$	0.2322±0.011			
		$K\beta_{2,4}/K\alpha$	0.0304±0.009			
		$K\beta_{2,4}/K\beta_{1,3}$	0.1693±0.009			
		$K\beta/K\alpha$	**0.1738±0.011**			
YBr₃	B=0	$K\beta_{1,3}/K\alpha$	0.2359±0.008			
		$K\beta_{2,4}/K\alpha$	0.0341±0.007			
		$K\beta_{2,4}/K\beta_{1,3}$	0.2004±0.009			
		$K\beta/K\alpha$	**0.1848±0.010**			
	B=0.6T	$K\beta_{1,3}/K\alpha$	0.2342±0.008			
		$K\beta_{2,4}/K\alpha$	0.0321±0.009			
		$K\beta_{2,4}/K\beta_{1,3}$	0.1987±0.010			
		$K\beta/K\alpha$	0.1818±0.010			
	B=1.2T	$K\beta_{1,3}/K\alpha$	0.2321±0.009			
		$K\beta_{2,4}/K\alpha$	0.0303±0.011			
		$K\beta_{2,4}/K\beta_{1,3}$	0.1976±0.010			
		$K\beta/K\alpha$	0.1794±0.009			
Y₂O₃	B=0	$K\beta_{1,3}/K\alpha$	0.2364±0.011			
		$K\beta_{2,4}/K\alpha$	0.0349±0.010			
		$K\beta_{2,4}/K\beta_{1,3}$	0.2008±0.010			
		$K\beta/K\alpha$	0.1856±0.011			
	B=0.6T	$K\beta_{1,3}/K\alpha$	0.2351±0.008			
		$K\beta_{2,4}/K\alpha$	0.0340±0.008			
		$K\beta_{2,4}/K\beta_{1,3}$	0.1998±0.010			
		$K\beta/K\alpha$	0.1831±0.010			
	B=1.2T	$K\beta_{1,3}/K\alpha$	0.2302±0.010			
		$K\beta_{2,4}/K\alpha$	0.0307±0.010			
		$K\beta_{2,4}/K\beta_{1,3}$	0.1985±0.010			

Element	External Magnetic Field	Intensity Ratio	This Work	Scofield (1974a)	Manson &Kennedy (1974)	Ertuğral et al. (2007)
		$K\beta/Ka$	0.1824±0.009			
YF$_3$	B=0	$K\beta_{1,3}/Ka$	0.2371±0.008			
		$K\beta_{2,4}/Ka$	0.0352±0.009			
		$K\beta_{2,4}/K\beta_{1,3}$	0.2011±0.010			
		$K\beta/Ka$	0.1859±0.010			
	B=0.6T	$K\beta_{1,3}/Ka$	0.2338±0.006			
		$K\beta_{2,4}/Ka$	0.0330±0.008			
		$K\beta_{2,4}/K\beta_{1,3}$	0.2003±0.011			
		$K\beta/Ka$	0.1841±0.010			
	B=1.2T	$K\beta_{1,3}/Ka$	0.2307±0.009			
		$K\beta_{2,4}/Ka$	0.0325±0.010			
		$K\beta_{2,4}/K\beta_{1,3}$	0.1989±0.011			
		$K\beta/Ka$	0.1829±0.011			
Y(SO$_4$)$_3$.8 H$_2$O	B=0	$K\beta_{1,3}/Ka$	0.2380±0.007			
		$K\beta_{2,4}/Ka$	0.0355±0.012			
		$K\beta_{2,4}/K\beta_{1,3}$	0.2015±0.010			
		$K\beta/Ka$	0.1867±0.011			
	B=0.6T	$K\beta_{1,3}/Ka$	0.2323±0.008			
		$K\beta_{2,4}/Ka$	0.0332±0.012			
		$K\beta_{2,4}/K\beta_{1,3}$	0.2006±0.011			
		$K\beta/Ka$	0.1844±0.012			
	B=1.2T	$K\beta_{1,3}/Ka$	0.2311±0.007			
		$K\beta_{2,4}/Ka$	0.0318±0.012			
		$K\beta_{2,4}/K\beta_{1,3}$	0.2001±0.010			
		$K\beta/Ka$	0.1816±0.011			
Y$_2$S$_3$	B=0	$K\beta_{1,3}/Ka$	0.2385±0.012			
		$K\beta_{2,4}/Ka$	0.0359±0.009			
		$K\beta_{2,4}/K\beta_{1,3}$	0.2019±0.009			
		$K\beta/Ka$	0.1876±0.008			
	B=0.6T	$K\beta_{1,3}/Ka$	0.2354±0.007			
		$K\beta_{2,4}/Ka$	0.0335±0.010			
		$K\beta_{2,4}/K\beta_{1,3}$	0.2009±0.009			
		$K\beta/Ka$	0.1867±0.011			
	B=1.2T	$K\beta_{1,3}/Ka$	0.2332±0.006			
		$K\beta_{2,4}/Ka$	0.0301±0.007			
		$K\beta_{2,4}/K\beta_{1,3}$	0.2002±0.008			
		$K\beta/Ka$	0.1849±0.012			

Table 6. $K\beta_{1,3}/Ka$, $K\beta_{2,4}/Ka$, $K\beta_{2,4}/K\beta_{1,3}$ and $K\beta/Ka$ X-ray intensity ratios of pure Y their compounds.

When you look at the Table 6, a serious difference between the $K\beta_{1,3}/K\alpha$ experimental and theoretical values can be seen. This situation is mainly because of the limited resolution of the detector. The $K\alpha_1$ and $K\alpha_2$ X-ray components appear as one line. In the most general case, chemical speciation is preferably performed via the analysis of the $K\beta_{1,3}$ or $K\beta_{2,4}$ lines. These lines, emitted after transition of valance electrons are more sensitive to the chemical environment.

As can be seen from Table 6, the $K\beta/K\alpha$ ratios of Y in all Y compounds are in close agreement with the ratios of corresponding pure metals. The greatest increase of the $K\beta/K\alpha$ ratio has been observed for Y_2S_3. We found a general increase of the $K\beta/K\alpha$ intensity ratios for different compounds. This situation is more complex because the $K\beta/K\alpha$ intensity ratio is affected by the chemical bonding type, (ionic, metallic, covalent), the individual characteristics of the structure of molecules, complexes and crystals (polarity, valency and electronegativity of atoms, co-ordination number, ionicities of covalent bond etc.).

We found that the chemical effect on the $K\beta/K\alpha$ ratios for 4d elements is small but the dependence of the $K\beta_2/K\alpha$ ratios on the chemical environments is appreciable. This can be understood by the fact that in 4d elements the valance state consists of the 4d, 5s and 5p electrons and the influence of the chemical state on the $K\beta_{1,3}$ (3p→1s) X-ray emission is negligible. Yamoto et al. (1986) found similar results for compounds involving Tc isotopes and Mukoyama et al. (2000) found similar results theoretically for Mo and Tc compounds.

The overall error in the present measurements is estimated to be 3-8%. This error is attributed to the uncertainties in different parameters used to determine the $K\beta_1/K\alpha$, $K\beta_2/K\alpha$, $K\beta_2/K\beta_1$ and $K\beta/K\alpha$ values; such as, $I_0 G\varepsilon$ product (1.0-2.5%), in the absorption correction factor (0.3-1.5%), the error in the area evaluation under the $K\alpha$, $K\beta_1$, $K\beta_2$ and $K\beta$ X-ray peak (0.5-3.0%) and the other systematic errors (1.0-2.0%).

4. Conclusion

There has been increasing interest in chemical speciation of the elements in recent years which can be attributed to the great alterations in the chemical and biological properties of the elements depending on their oxidation state, the type of chemical bonds etc. Usually, the influence of the chemical environment results in energy shifts of the characteristic X-ray lines, formation of satellite lines and changes in the emission linewidths and relative X-ray intensities. High resolution X-ray spectroscopy, employing crystal spectrometers of a few eV resolutions, can be applied to probe these phenomena efficiently, exploiting them for chemical state analysis. Measurements of the shapes and wavelengths of certain X-ray lines have been made by previous investigators with both EDXRF and WDXRF. It has been shown that the WDXRF spectrometer is capable of measuring X-ray wavelengths with a precision equal to or greater than that attained with the EDXRF system. Both EDXRF and WDXRF technique has been used to study the effect of chemical state of an element on characteristic X-rays.

We have presented and discussed the effect of chemical composition and external magnetic field on the $K\beta_{1,3}/K\alpha$, $K\beta_{2,4}/K\alpha$, $K\beta_{2,4}/K\beta_{1,3}$ and $K\beta/K\alpha$ intensity ratios for some Yttrium compounds. The experimental measurements have been performed with a Si(Li) detector. The observed spectral features, namely the asymmetry indices, FWHM values, chemical shifts, energy separations between $K\alpha$ and $K\beta$ lines and $K\beta/K\alpha$ intensity ratio values show an interesting correlation with crystal symmetries. Furthermore, these values change symmetrically with the external magnetic field. There is a relation between the crystal structures and K X-ray emission rate because of the change in bond distance, inter atomic distance, the interaction between ligand atoms and the central atom, and the Auger electron and dipole transition. These situations cause a redistribution of the electron configuration in the molecule.

A correlation between the $K\beta/K\alpha$ intensity ratio of 4d elements and chemical state was found in this work. Excluding the values for Y, we can generally state that $K\beta/K\alpha$ intensity ratio increases for different compounds. The $K\beta_1/K\alpha$, $K\beta_2/K\alpha$, $K\beta_2/K\beta_1$ and $K\beta/K\alpha$ intensity ratio values were obtained in the present work and listed in Table 6 and compared with other experimental and theoretical values. As a result, we can say that the uncertainties of the measured values are too large to allow any statement about the specific dependence of the $K\beta/K\alpha$ intensity ratio on the crystal symmetry, but small enough to show significant increase in the $K\beta/K\alpha$ intensity ratio with increasing external magnetic field values.

In general, our experimental values are qualitatively in agreement with the other experimental values. There are some differences between the results of this study and that of previous experimental work because these studies were carried out in different laboratories and different systems. We were not obtained researches interested in $K\beta_{1,3}/K\alpha$, $K\beta_{2,4}/K\alpha$, $K\beta_{2,4}/K\beta_{1,3}$ and $K\beta/K\alpha$ intensity ratio values for Y compounds. So we do not compare compounds these intensity ratio values in literature values. Rigorous systematic experiments and theoretical calculations are urgently needed for comparison with present experimental result. To obtain more definite conclusions on the magnetic field and crystal structure dependency of the atomic parameters, more experimental data are clearly needed, particularly for different symmetries and for chemical compounds.

5. Acknowledgment

This work was supported by the Scientific and Technological Research Council of Turkey (TUBITAK), under the project no 106T045.

6. References

Allinson, S.K. (1933). The Natural Widths of the $K\alpha$ X-Ray Doublet from [26]Fe to [47]Ag. *Phys. Rev.* Vol.44, pp. 63-72.

Arndt, E.; Brunner, G. & Hartmann, E. (1982). $K\beta/K\alpha$ Intensity Ratios for 3d Elements by Using Photoionisation and Electron Capture. *J. Phys. B: At. Mol. Opt. Phys.*, Vol.15, pp. 887-889.

Balasubramanian, K. (1994). Relativistic Effects and Electronic Structure of Lanthanide and Actinide Molecules, in: Gschneidner, K.A.; Eyring, L.; Choppin, G.R. & Lander G.H. (Ed.), *Handbook of Physics and Chemistry of Rare Earth*, 18, Elsevier, Amsterdam, Chap. 119, pp. 29-50.

Band, I.M.; Kovtun, A.P.; Listengarten, M.A. & Trzhaskovskaya, M.B. (1985). The Effect of the Chemical Environment of Manganese and Chromium Atoms on the $K\beta/K\alpha$ X-Ray Intensity Ratio. *J. Electr. Spectr. and Relat. Phenom.*, Vol.36, pp. 59-68.

Batrakov, Y.F.; Krivitsky, A.G. & Puchkova E.V. (2004). Relativistic Component of Chemical
Shift of Uranium X-Ray Emission Lines. *Spectrochim. Acta Part B*, Vol.59, pp. 345
351.

Berenyi, D.; Hock, G.; Ricz, S.; Schlenk, B. & Valek, A. (1978). $Ka/K\beta$ X-Ray Intensity Ratios
and K-Shell Ionisation Cross Sections for Bombardment by Electrons of 300-600
keV. *J. Phys. B*, Vol.11, pp. 709-713.

Berger, M.J. & Hubbell, J.H. (1987). XCOM: Photon Cross-Sections on a Personnel Computer
with (Version 1.2).

Brunner, G.; Nagel, M.; Hartmann, E. & Arndt, E. (1982). Chemical Sensitivity of
the $K\beta/Ka$ X-Ray Intensity Ratio for 3d Elements. *J. Phys. B*, Vol.15, pp. 4517-
4522.

Demir, D. & Şahin, Y. (2006a). Measurement of the K shell X-Ray Production Cross-
Sections and Fluorescence Yields for Nd, Eu, Gd, Dy and Ho using Radioisotopes
X-Ray Fluorescence in the External Magnetic Field. *Eur. Phys. J. D*, Vol.44, pp. 34-
38.

Demir, D. & Şahin, Y. (2006b). The Effect of an External Magnetic Field on the L_3 Subshell
Fluorescence Yields and Level Widths for Gd, Dy, Hg and Pb at 59,54 keV. *Nucl.
Instr. and Meth.*, Vol.254, pp. 43-48.

Deutsch, M. & Hart, M. (1982). Wavelength, Energy Shape, and Structure of the Cu Ka_1 X
Ray-Emission Line. *Phys. Rev. B*, Vol.26, pp. 5558-5567.

Ertuğral, B.; Apaydın, G.; Çevik, U.; Ertuğrul, M. & Kopya, A.İ. (2007). $K\beta/Ka$ X-Ray
Intensity Ratios for Elements in the Range 16<Z<92 Excited by 5.9, 59.5 and 123.6
keV Photons. *Radiation Phys. and Chem.*, Vol.76, pp. 15-22.

Fichter, M. (1975). Das K-Röntgenemissionsspektrum von Phosphor in Abhängigkeit von
der Chemischen Bindung. *Spectrochim. Acta Part B*, Vol.30, pp.417-431.

Giauque, R.D.; Goulding, F.S.; Jaklevic, J.M. & Pehl, R.H. (1973). Trace Element
Determination with Semiconductor Detector X-Ray Spectrometers. *Anal. Chem.*,
Vol.45, pp. 671-681.

Gohshi, Y. & Ohtsuka, A. (1973). The Application of Chemical Effects in High Resolution X-
Ray Spectrometry. *Adv. X-Ray Anal.*, Vol. 28, pp. 179-188.

Gohshi, Y.; Hirao, O. & Suzuki, I. (1975). Chemical State Analysis of Sulfur, Chromium
and Tin by High Resolution X-Ray Spectrometry. *Adv. X-Ray Anal.*, Vol.18, pp.
406-414.

Groot, F.M.F. (1994a). X-Ray Absorption and Dichroism of Transition Metals and Their
Compounds. *J. Electron Spectrosc. Relat. Phenom.*, Vol.676, pp. 529-622.

Groot, F.M.F. (1994b). New Directions in Research with Third-Generation Soft X-Ray
Synchrotron Radiation Sources. *Applied Sciences*, Vol.254.

Han, I.; Porikli, S.; Şahin, M. & Demir. D. (2010). Measurementof La, $L\beta$ and Total L X-ray
Fluorescence Cross-Sections for Some Elements with 40<Z<53. *Radiation Physics and
Chemistry*, Vol.79, pp. 393–396.

Iiahara, J.; Omorr, T., Yoshihara, K.; & Ishii, K. (1993). Chemical Effects on Chromium L X-
Rays. *Nucl. Instr. Methods. B*, Vol.73, pp. 32-34.

Kataria, S.K.; Govil, R.; Saxena, A. & Bajpei, H.N. (1986). Chemical Effects in X-ray
Fluorescence Analysis. *X-Ray Spectr.*, Vol.15, pp. 49-53.

Katz, J.J.; Seaborg, G.T. & Morss, L.R. (1986). 2nd ed, The Chemistry of the Actinide Elements, 2, *Chapman and Hall*, New York, pp. 1131–1165.

Khan, M.R. & Karimi, M. (1980). $K\beta/K\alpha$ Ratios in Energy-Dispersive X-Ray Emission Analysis. *X-Ray Spectrom.*, Vol.9, pp. 32-35.

Krause, M.O. (1979). Atomic Radiative and Radiationless Yields for K and L Shells. *Chem. Ref. Data.*, Vol.8, pp. 307-327.

Küçükönder, A.; Şahin, Y.; Büyükkasap, E. & Kopya, A. (1993). Chemical effect on $K\beta/K\alpha$ X Ray Intensity Ratios in Coordination Compounds of Some 3d Elements. *J. Phys. B: At. Mol. Opt. Phys.*, Vol.26, pp. 101-105

Leonhardt, G. & Meisel, A. (1970). Determination of Effective Atomic Changes From the Chemical Shifts of X-Ray Emission Lines. *J. Chem. Phys.* Vol.52, pp. 6189-6198.

Makarov, L.L. (1999). X-ray Emission Effects as a Tool to Study Light Actinides. *Czech. J. Phys.*, Vol.49, pp. 610–616.

Manson, S. T. (1974). X-Ray Emission Rates in the Hartree–Slater Approximation. *At. Data Nucl. Data Tables.* Vol.14, pp. 111-120.

Meisel, A.; Leonhardt, G. & Szargan, R. (1989). X-Ray Spectra and Chemical Binding, *Chemical Physics*, Vol.37, edited by F.P. Schafer, V.I. Goldanskii & J.P. Toennies (Springer-Verlag, Berlin).

Mukoyoma, T.; Taniguchi, K. & Adachi, H. (1986). Chemical Effect on $K\beta/K\alpha$ X-Ray Intensity Ratios. *Phys. Rev. B*, Vol.34, pp. 3710-3716.

Mukoyoma, T.; Taniguchi, K. & Adachi, H. (2000). Variation of $K\beta/K\alpha$ X-Ray Intensity Ratios in 3d elements. *X-Ray Spectr.*, Vol.29, pp. 426-429.

Mukoyama, T. (2004). Theory of X-ray Absorption and Emission Spectra. *Spectrochim. Acta Part B*, Vol.59, pp. 1107-1115.

Nagel, D.J. & Baun, W.L. (1974). In: L.V. Azaroff (Ed.), *X-Ray Spectroscopy*, McGraw-Hill, US, Ch. 9.

Padhi , H.C.; Bhuinya, C.R. & Dhal, B.B. (1993). Influence of Solid-State Effects on the $K\beta/K\alpha$ Intensity Ratios of Ti and V in TiB_2, VB_2 and VN. *J. Phys. B: At. Mol. Opt. Phys.*, Vol.26, pp. 4465-4469.

Padhi , H.C. & Dhal, B.B. (1995). $K\beta/K\alpha$ X-Ray-Intensity Ratios of Fe, Co, Ni, Cu, Mo, Ru, Rh and Pd in Equiatomic Aluminides. *Solid State Commun.*, Vol. 96, pp. 171-173.

Paic, G. & Pecar, V. (1976). Study of Anomalies in $K\beta/K\alpha$ Ratios Observed Following K Electron Capture. *Phys. Rev. A*, Vol.14, pp. 2190-2192.

Pepper, M. & Bursten, B.E. (1991). The Electronic Structure of Actinide Containing Molecules: a Challenge to Applied Quantum Chemistry. *Chem. Rev.*, Vol.91, pp. 719–741.

Porikli, S. & Kurucu, Y. (2008a). The Effect of an External Magnetic Field on the $K\alpha$ and $K\beta$ X-Ray Emission Lines of the 3d Transition Metals. *Instr. Science and Tech.*, Vol.36:4, pp. 341-354.

Porikli, S. & Kurucu, Y. (2008b). Effects of the External Magnetic Field and Chemical Combination on $K\beta/K\alpha$ X-Ray Intensity Ratios of Some Nickel and Cobalt Compounds. *Appl. Rad. and Isot.*, Vol.66, pp. 1381-1386.

Determination of Chemical State and External Magnetic Field Effect on the Energy Shifts and X-Ray Intensity
Ratios of Yttrium and Its Compounds

109

Porikli, S.; Demir, D. & Kurucu, Y. (2008c). Variation of $K\beta/Ka$ X-Ray Intensity Ratio and Lineshape with the Effects of External Magnetic Field and Chemical Combination. *Eur. Phys. J. D*, Vol.47, pp. 315–323.

Porikli, S.; Han, İ.; Yalçın, P. & Kurucu, Y. (2011a). Determination of Chemical Effect on the $K\beta_1/Ka$, $K\beta_2/Ka$, $K\beta_2/K\beta_1$ and $K\beta/Ka$ X-Ray Intensity Ratios of 4d Transition Metals. *Spectroscopy Letters*, Vol.44, pp. 38–46.

Porikli, S. (2011b). Influence of the Chemical Environment Changes on the Line Shape and Intensity Ratio Values for La, Ce and Pr L Lines *Spectra. Chem. Phys. Lett.*, Vol.508, pp. 165-170.DOI information: 10.1016/j.cplett.2011.04.021

Pyykko, P. (1988). Relativistic Effects in Structural Chemistry. *Chem. Rev.*, Vol.88, pp. 563-94.

Raj, S.; Padhi, H.C. & Polasik, M. (1998). Influence of Chemical Effect on the $K\beta$-to-Ka X-Ray Intensity ratios of Ti, V, Cr, and Fe in TiC, VC, CrB, CrB_2 and FeB. *Nucl. Instrum. Meth. in Phys. Res. B*, Vol.145, pp. 485–491.

Rao, N.V.; Reddy, S.B; Satyanarayana, G. & Sastry, D.L. (1986). $K\beta/Ka$ X-Ray Intensity Ratios. *Physica C*, Vol.142, pp. 375-380.

Salem, S.I.; Panossian, S.L. & Krause, R.A. (1974). Quantum Mechanics of Atomic Spectra and Atomic Structure. *At. Data Nucl. Data Tables*, Vol.14, pp.91-109.

Scofield, J.H. (1969). Radiative Decay Rates of Vacancies in K and L Shells. *Phys. Rev.*, Vol.179, pp. 9-16.

Scofield, J.H. (1973). Theoretical Photoionization Cross Sections from 1 to 1500 keV, Unpublished *Lawrence Livermore Laboratory Report* UCRL-51326, Livermore, California

Scofield, J.H. (1974a). Relativistic Hartree–Slater Values for K and L Shell X-Ray Emission Rates. *Atomic Data and Nuclear Data Tables*, Vol.14, pp. 121-137.

Scofield, J.H. (1974b). Exchange Corrections of K X-ray Emission Rates. *Phys. Rev. A*, Vol.9, pp. 1041–1049.

Stöhr, Y. & Wu, Y. (1994). X-Ray Magnetic Circular Dichroism: Basic Concepts and Theory for 3d Transition Metal Atoms. *New Directions in Research with 3rd Generation Soft X-Ray Synchrotron Radiation Sources*, Editors F. Schlachter and F. Wuilleumier (Kluwer, Netherlands, 1993), p. 221.

Tamaki, Y.; Omori, T. & Shiokawa, T. (1979). Chemical Effect on the $K\beta/Ka$ Intensity Ratios of the Daughter Atoms Formed by the EC decay of ^{51}Cr and [54]Mn. *Radiochem. Radioanal. Lett.*, Vol.37, pp. 39-44.

Taniguchi, K. (1984). Chemical-State Analysis by Means of Soft X-Ray Spectroscopy. 2. $K\beta$ Spectra for Phosphorus, Sulfur, and Chlorine in Various Compounds. *Bull. Chem. Soc. Jpn.*, Vol.57, pp. 915-920.

Thole, B.T.; Carra, P.; Sette, F. & Van der Lean, G. (1992). X-Ray Circular-Dichroism as a Probe of Orbital Magnetization. *Phys. Rev. Lett.*, Vol.68, pp. 1943-1946.

Torres Deluigi, M.; Perino, E.; Olsina R. & Riveros de la Vega, A. (2003). Sulfur and Phosphorus $K\beta$ Spectra Analyses in Sulfite, Sulfate and Phosphate Compounds by X-Ray Fluorescence Spectrometry. *Spectrochim. Acta Part B*, Vol.58, pp- 1699-1707.

Urch, D.S. (1979). Theory, Techniques, and Application; Brundle, C. R., Baker, A. D., Eds.; Academic Press: New York, *Electron Spectrosc.*, Vol.3, pp. 1-39.

Yamoto, I.; Kaji, H. & Yoshihara, K. (1986). Studies on Chemical Effects on X-Ray Intensity Ratios of $K\beta/Ka$ in Nuclear Decay of Technetium Nuclides ^{99m}Tc, ^{97m}Tc and ^{95m}Tc. *J. Chem. Phys.*, Vol.84, pp- 522-527.

6

Application of Enriched Stable Isotopes in Element Uptake and Translocation in Plant

Shinsuke Mori[1], Akira Kawasaki[2], Satoru Ishikawa[2] and Tomohito Arao[2]
[1]NARO Western Region Agricultural Research Center
[2]National Institute for Agro-Environmental Sciences
Japan

1. Introduction

Isotope technique including radioisotopes and stable isotopes is useful and potent tool for various scientific areas. Especially, enriched stable isotopes are indispensable tools for researchers in biological systems (Stürup et al. 2008).

Stable isotope ratios are usually used in examining the biogeochemical cycling of light elements such as carbon(C), oxygen (O), nitrogen (N) and sulphur (S) in the environment. Thermal ionization mass spectrometry (TIMS) for the isotope analysis has been the most standard technique for many years. However, for TIMS analysis, time for sample preparation is needed because sample need to ensure efficient ionization. On the other hand, ICP-MS analysis has some advantages that sample preparation is simple and high sample throughput for isotope experiments where a large amount of samples need to be analyzed (Stürup et al. 2008). The disadvantage to resolve in isotope analysis using ICP-MS is spectroscopic interferences in the process of analysis. It is therefore needed to be resolved these interferences.

When plant physiologists investigate mineral absorption mechanisms in roots of plant, evaluation of symplastic mineral absorption capacity in roots cell in kinetics and time course experiments is very important because mineral translocation in shoots is mainly contributed to capacity of symplastic absorption in roots. In these experiments, radioisotopes methods are mainly used for element uptake in plants. Radioisotopes in solute were the most useful markers used in nutrient uptake and translocation in plants because they are chemically similar to the solute and can be distinguished from non-labeled solutes already contained in the roots (Davenport 2007). However, there are limitations to this method, including radioisotope administrative restriction and the restricted half-life of the radioisotope. Isotope tracer experiments, using a stable isotope, are very similar to those using a radioisotope on element to analyse plant mechanisms (Stürup et al. 2008). Accurate and precise determination of mineral isotope ratios is required for analysis of enriched stable isotopes. Inductive coupled plasma mass spectrometry (ICP-MS) has now become the effective and potent technique for enriched stable isotope tracer experiments due to increased availability. Therefore, the application of enriched stable isotopes in various biological systems increased rapidly.

There are so many research using enriched stable isotopes used as tracers aquatic and terrestrial ecosystems, animals and humans (See review of Stürup et al. 2008). However, there are a few researches using enriched stable isotopes element in plants. Recently, Stürup et al. (2008) reviewed that application of enriched stable isotopes as tracers in biological systems including aquatic ecosystem, terrestrial ecosystem, animals and humans in detail. Therefore, we did not focus on aquatic ecosystem animal and human in this chapter.

In this chapter, we therefore provide a review of some example using isotope technique. Especially, we focus on the application of enriched stable isotopes element uptake and translocation in plants. Our new method for evaluation of symplastic absorption of roots introduced in Section 4 has some merits, compared to radioisotopes techniques. Application of stable isotopes will become a new tool to evaluate element behavior in plants.

2. Application of stable isotopes in plants

The biochemical cycling of light element such as carbon(C), oxygen(O), nitrogen(N) and sulphur(S) have been studying using stable isotopes. The mechanisms of photosynthesis and of element uptake and translocation in plants was clarified by these studies using stable isotopes ratios such as C,O,N and S. Recently, the application of enriched isotopes of such as Mg, Cu, Ca, K and Cd behavior in plants rapidly increased with the development of ICP-MS analysis techniques. There are several studies on element uptake and translocation in plant using enriched stable isotopes (Table1).

Isotopes	Aim of study and method	Reference
^{10}B, ^{11}B	Characterization of boron uptake and translocation in sunflower plant. After preculture under nutrient solution containing ^{11}B,a short time experiment were conducted under nutrient solution containing low or high ^{11}B.	Dannel et al. (2000)
^{10}B	After preculture grown in nutrient solution containing boron, uptake experiment was conducted in solution containing enriched stable isotopes of ^{10}B.	Takano et al. (2002)
^{113}Cd	Intact leaves and cell sap of Cd accumulator plant were subjected to ^{113}Cd-NMR and H-NMR analysis for identification of the form of Cd in leaves.	Ueno et al. (2005)
^{113}Cd and ^{114}Cd	To examine Cd uptake in roots of *solanum* species with different Cd accumulation in shoot, uptake experiments were conducted using ^{113}Cd and ^{114}Cd.	Mori et al. (2009b)
^{113}Cd	Cd accumulation stage in soybean seed was examined in hydroponic solution using enriched isotope of ^{113}Cd.	Yada et al. (2004), Oda et al. (2004)
^{113}Cd	Cd uptake mechanisms in soybean was examined using ^{113}Cd isotopes in pot and field experiment	Kawasaki et al. (2004,2005)

Table 1. Element uptake and translocation in plant using enriched stable isotopes

Dannel et al. (2000) characterized the boron uptake and translocation from roots to shoots in sunflower using the stable isotopes ^{10}B and ^{11}B. In the report, after sunflower plant was precultured with high(100 μM) or low (1 μM) ^{11}B supply, plants were treated under differential ^{10}B supply condition. The results suggested that B uptakes are mediated by two transport mechanisms. First mechanism is passive diffusion which is indicated by the linear components. Second mechanism is energy dependent process which is indicated by the saturated components. Kawasaki et al. (2004, 2005) conducted that an isotope tracer technique with ^{113}Cd has been used in pot and field experiments. They examined that the most critical stages of soybean in which Cd absorbed via roots was transferred into the seeds. Cd absorbed before the beginning seed stage causes an increase of Cd concentration in seeds. Yada et al.(2004) reported that soybean plants were grown in hydroponic solution and supplied ^{113}Cd via roots for 48 h at early growth stage to investigate Cd accumulation pathway in soybean seed using enriched isotope of ^{113}Cd. Cd accumulated in leaves was translocated to seeds at seed beginning maturity stage. Oda et al. (2004) also indicated that the Cd absorbed from full pod to full seed was the most contributive to raise the Cd amount of seeds. Ueno et al. (2005) reported that *Thlaspi caerulescens* which is Cd hyperaccumulator plants have been grown hydroponically with a highly enriched ^{113}Cd isotope to investigate the form of Cd in the leaves using ^{113}Cd nuclear magnetic resonance (NMR) spectroscopy. They identified that cadmium binds with malate in the leaves. Several enriched isotopes such as ^{111}Cd, ^{113}Cd and ^{114}Cd will become a new tool to evaluate Cd behavior in plants. Several studies stated above suggest that enriched isotope is a very potent technique for tracking the distribution, uptake, translocation and recycling in biological system. Now, many enriched element stable isotopes except B and Cd are able to purchase in chemical forms such as metallic or oxide. In the future, the benefit of enriched stable isotopes techniques would be paid much attention in plant and environmental science areas.

3. Several methods for evaluating symplastic element uptake in plants

Intensive studies on the absorption mechanisms of various elements by plant roots have been conducted. There are evidence on mineral uptake and translocation in plants. It is well known that ion absorption in plant roots shows a saturated curve in kinetics experiments, indicating that a type of proteinaceous transporter mediates ion absorption (Epstein and Hagen 1952). Plant physiologists examining ion absorption in plant roots have given much attention to ion transport via the symplast across the plasma membrane (Epstein 1973). However, when ion absorption experiments were conducted, it was found that the apoplastically absorbed ions needed to be washed out of the apoplast to determine the symplastically absorbed ions across the plasma membrane or the determination of absorption is overestimated (Glass 2007). Therefore, it is necessary to eliminate the apoplastically bound ions to evaluate the symplastically absorbed ion content in the roots. To evaluate symplastic cadmium(Cd) and other elements absorption in roots, several methods have generally been used in the past: (1) expose the plant material to Cd radioisotopes and subsequent desorption using unlabelled Cd in the root apoplast (Hart et al. 1998, 2002, 2006), (2) plant material is exposed to Cd radioisotopes under conditions at 2°C and 22°C (Zhao et al. 2002, Uraguchi et al. 2009), (3) metabolic inhibitors such as DNP or CCCP (Cataldo et al.1983, Ueno et al. 2009), (4) centrifuge method (Yu et al. 1999, Mitani and Ma 2005, Ma et al. 2004, Ueno et al. 2008), (5)estimation of desorption from roots with time(Lasat et al. 1998)

Regarding evaluation for symplastic element uptake in roots using radioisotopes, this method is used for symplastic element uptake in roots. Hart et al. (1998, 2002, 2006) reported that Cd uptake experiment was conducted in nutrient solution containing [109]Cd-labeled $CdSO_4$ and apoplastic [109]Cd were desorbed using excessive nonlabelled Cd. As other method, Nakanishi et al. (2006) evaluated that apoplastic Cd in the roots was washed in 0.5 mmol L[-1] ethylenediaminefetraacetic acid (EDTA) for 1 min. Lasat et al. (1996) evaluated that symplastic Zn uptake in roots of Zn hyperaccumulator and nonaccumulator *Thlaspi* species apoplastic [65]Zn in roots desorbed by excessive unlabelled $ZnCl_2$ solution after Zn uptake experiment was conducted using [65]Zn raidoisotopes. There is merit that this method is able to detect radioisotope element with high sensitivity. However, there are limitations to this method, including radioisotope administrative restriction and restricted half of the radioisotope. Additionally, the radioisotope technique has toxicological concern. It is required for handling its isotopes to be careful.

Regarding evaluation of symplastic element uptake in roots using differences in the amounts of Cd absorbed at 2°C and 25°C. Uptake of element at 2°Cwas assumed to represent mainly apoplastic binding in the roots whereas the difference in uptake between 22°C and 2°C represented metabolically dependent influx. Zhao et al. (2002) reported that apoplastic and symplastic uptake in two *Thlaspi* species from Cd and Zn depletion in solution using radioisotope tracer. Uraguchi et al.(2009) reported that genotypic variation in cadmium accumulation in rice and evaluated that symplastic Cd uptake in roots of rice using the method of subtraction the Cd content in the roots at 2°C from the Cd content in the roots at 25°C. This method using unlabeled Cd is easy to handle because there is no administrative limitation not using radioisotope elements. However, this method needs double seedlings for evaluation. Additionally, this method cannot be evaluated using same seedling. This method is not easy for dicotyledonous plant such as *Solanum melongena* to handle.

As for methods using metabolic inhibitors, Cataldo et al. (1983) reported that Cd uptake dependent on energy in roots is suppressed by dinitrophenol as metabolic inhibitor. In this study, using dinitrophenol as a metabolic inhibitor, the 'metabolically absorbed' fraction was shown to represent 75 to 80% of the total absorbed fraction at concentration less than 0.5 μ mol, and decreased to 55% at 5 μ mol.

Regarding centrifuge method, tap roots of plants were harvested and 2 cm root tips were excised. Then, cut ends were washed in distilled water and blotted dry. For each sample, 30 roots were used. The cut ends were washed in distilled water quickly and blotted dry. The tips were placed in a 0.45 mM filter unit with the cut ends facing down and centrifuged at 2,000g for 15 min at 4°C to obtain the apoplastic solution. After centrifugation, root segments were frozen at -80°C for 2 h and then thawed at room temperature. The symplastic solution was prepared from frozen-thawed tissues by centrifugation at 2,000g for 15 min at 4°C. Ma et al. (2004) evaluated that symplastic Si uptake of wild type rice and mutant rice using this centrifuge method. Additionally, Mitani and Ma (2005) also evaluated that symplastic Silicon uptake in rice, tomato and cucumber which differ from Si accumulation capacity using this method. Ueno et al.(2009) reported that symplastic Cd uptake is estimated by cell sap obtained from centrifuge method. To check the purity of apoplastic solution, the activity of malic dehydrogenase in apoplastic and symplastic solution was determined. The activity of malic dehydrogenase in apoplastic solution was below one-twentieth and approximately one-fortieth of symplastic solution. This method is valuable for evaluation of symplastic Cd concentration in roots because Cd concentration in roots cell cap was directly determined. However, evaluation using root tips possibly is not representative of most root tissues. Rain et al.(2006) pointed out that there are the difference of K_m value in kinetics experiment between whole roots and root tips.

As other evaluation method of roots fraction, Lasat et al.(1998) evaluated that each fraction of cell wall, cytoplasm and vacuole by each efflux fraction from roots. They investigated that difference of Zn fraction in roots such as cell wall, cytoplasm and vacuole using this method.

4. Application of enriched stable isotopes in element uptake and translocation in plant

In this section, we introduce that our new method for evaluation of symplastic ion absorption, especially cadmium (Mori et al. 2009a). Several methods stated above is evaluation that apoplastically bound element is desorbed by some elements after element absorption experiment. Our method is that symplastic Cd absorption capacity is evaluated by difference of enriched isotope of ^{113}Cd and ^{114}Cd. Cadmium (Cd) is a hazardous heavy metal with regards to human health and is dispersed in natural and agricultural environments principally through human activities (Wanger, 1993). Arable land contains, to some extent, Cd, reportedly in the range, 0.04–0.32M, even in non-polluted soil (Keller, 1995; Wanger, 1993). This results in Cd accumulation in the edible parts of crops. Recently, the Codex Alimentarius Commission (2005) adopted a maximum concentration of 0.05 mg Cd kg^{-1} (fresh weight) recommended for fruiting vegetables. Approximately 7% of 381 samples of eggplant (*Solanum melongena*), 22% of 165 samples of okra (*Abelmoschus esculentus*), and 10% of 302 samples of taro (*Colocasia esculenta*) contained Cd concentrations above this limit in a field and market-basket study during 1998–2001 in Japan (Ministry of Agriculture Forestry and Fisheries of Japan, 2002); despite the fact that these crops were cultivated in non-polluted fields. Under these circumstances, new technologies for reducing the Cd level in crops are urgently required in Japan. Therefore, it is important to elucidate the mechanisms mediating Cd absorption, accumulation, and translocation in these crops. The crop conditions were represented by low Cd concentration experimental mediums.

4.1 Validity of our method for evaluation of symplastic Cd uptake in roots using enriched isotopes of ^{113}Cd and ^{114}Cd

When ion absorption experiments were conducted, it was found that the apoplastically absorbed ions needed to be washed out of the apoplast to determine the symplastically absorbed ions across the plasma membrane or the determination of absorption is overestimated(Glass 2007). Therefore, it is necessary to eliminate the apoplastically bound ions to evaluate the symplastically absorbed ion content in the roots. There are several methods to eliminate apoplastic ions as stated above. In this section, we introduced our new method for symplastic Cd absorption in roots of *Solanum melongena* using enriched isotopes of ^{113}Cd and ^{114}Cd.

The enriched isotopes of ^{113}Cd (^{106}Cd, 0.16%; ^{108}Cd, 0.135%; ^{110}Cd, 0.81%; ^{111}Cd, 2.53%; ^{112}Cd, 2.61%; ^{113}Cd, 93.29%; ^{114}Cd, 0.46%; ^{116}Cd, 0.01%) and ^{114}Cd (^{106}Cd, 0.05%; ^{108}Cd, 0.05%; ^{110}Cd, 0.05%; ^{111}Cd,0.05%; ^{112}Cd, 0.05%; ^{113}Cd, 5.6%; ^{114}Cd, 93.6%; ^{116}Cd,0.8%) used in the present study were purchased from Isoflex (San Francisco, CA,USA) in metallic form and dissolved in diluted HNO_3. The enriched isotopes of ^{114}Cd contained the 5.6 % of ^{113}Cd.

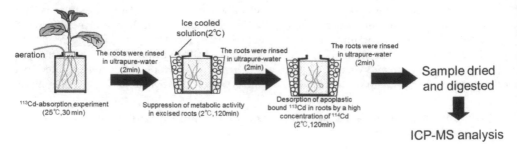

(modified from Mori et al. 2009a)

Fig. 1. Absorption experiment procedure for evaluating symplastic ^{113}Cd absorption in roots

The procedure for evaluating symplastic Cd absorption in the roots, using enriched isotopes ^{113}Cd and ^{114}Cd, is illustrated in Fig. 1. The roots of intact seedlings were rinsed in ultrapure water for 2 min and then exposed to a 500 mL ^{113}Cd solution containing 0.5 mmol L^{-1} CaCl$_2$ and 2 mmol L^{-1} 2-morpholinoethanesulfonic acid monohydrate Tris (hydroxymethyl) aminomethane (MES–Tris) (pH 6.0) at 25°C for 30 min (Fig. 1). The levels of ^{113}Cd were 40 nmol or 400 nmol in the ^{113}Cd treatment. A-B shown in Fig.2 indicates that ^{113}Cd absorbed in roots consists of apoplastic ^{113}Cd and symplastic ^{113}Cd (Fig.2 A, B). To suppress metabolically dependent symplastic absorption from the apoplast, the roots were excised from each seedling and immersed in a cold Cd-free buffer solution (2 mmol L^{-1} MES–Tris [pH 6.0], 0.5 mmol L^{-1} CaCl$_2$) at 2°C for 120 min (Fig. 1, Fig.2 C). The apoplastic-bound ^{113}Cd in the roots from 40 or 400 nmol ^{113}Cd treatment was then desorbed by immersing the roots in the same cold buffer solution at 2°C containing a 50-fold concentration of ^{114}Cd (2 or 20 μmol) for 120 min (Fig.1, Fig.2 D, E, F). The excised roots were then rinsed in ultrapure water for 2 min. Harvested samples were dried in an oven at 75°C for 3 days until dry. After digestion of dried sample, we then determined ^{113}Cd and ^{114}Cd contents in roots by ICP-MS analysis. To confirm the validity of this method, we compared our Cd absorption results with the Cd absorption results obtained at 25°C and 2°C using unlabeled CdCl$_2$ reagent. The experimental procedure was as follows. The Cd-absorption experiments were conducted for 30 min using 500 mL solutions containing 2 mmol L^{-1} MES–Tris (pH 6.0), 0.5 mmol L^{-1} CaCl$_2$ and different concentrations of Cd (40 or 400 nmol) at 25°C. After the absorption experiment, the excised roots from each seedling were rinsed with ultrapure water for 2 min. For the Cd-absorption experiment at 2°C, plants were transferred to an ice-cold pretreatment solution containing 2 mmol L^{-1} MES–Tris (pH 6.0) and 0.5 mmol L^{-1} CaCl$_2$ for 120 min. The Cd-absorption experiment at 2°C was conducted for 30 min. In the unlabeled Cd-absorption experiment at different temperatures, the amount of Cd reportedly absorbed into roots at 2°C was estimated to be apoplastically bound Cd on the assumption that metabolically dependent absorption would be suppressed at low temperature. Therefore, the difference in the amount of Cd absorbed at 2°C and at 25°C represents symplastic Cd absorption depending on metabolic energy. All absorption experiments were replicated three times. Each procedure illustrated in Figure1 signifies a schematic representation shown in Fig. 2.

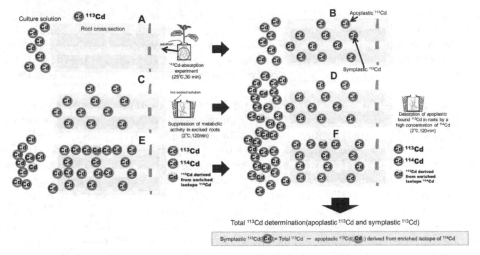

Fig. 2. A schematic representation of Cd absorption and desorption in roots using different enriched isotopes

4.2 Determination of ^{113}Cd, ^{114}Cd and the Cd contents in the roots

Approximately 0.05–0.1 g of dried roots was transferred and digested in a 10 mL Teflon tube containing 3 mL HNO_3. After digestion, the digested solution was diluted and 10 ng mL^{-1} of indium (In) was added to each diluted solution as an internal standard for ^{114}Cd determination. For ^{113}Cd determination, 10 ng mL^{-1} of tellurium (Te) was added as an internal standard. The concentrations of ^{113}Cd and ^{114}Cd in the digested solutions were determined by ICP-MS (ELAN DRC-e; Perkin Elmer SCIEX, Concord, ON, Canada). The concentrations of Cd in the digested solutions from the Cd-absorption experiment using unlabeled $CdCl_2$ reagent were determined by ICP atomic emission spectroscopy (VISTA-PRO; Varian, Palo Alto, CA, USA). It is well known that MoO interferes spectroscopically in determining the concentration of Cd in ICP-MS analysis (Kimura et al. 2003; May and Wiedmeyer 1998). In addition, it has been shown that it is necessary to remove Mo from the digested solution to avoid spectroscopic interference by molybdenum oxides (Oda et al. 2004; Yada et al. 2004). Therefore, for the ^{113}Cd and ^{114}Cd count intensities, we monitored the spectroscopic interference of the molybdenum oxides (^{97}Mo^{16}O and ^{98}Mo^{16}O) detected in the 10 ng mL^{-1} Mo standard solution. The contribution rate of spectroscopic interference of the putative ^{97}Mo^{16}O and ^{98}Mo^{16}O for ^{113}Cd and ^{114}Cd contents was negligibly small in both treatments (40 and 400 nmol). Therefore, we considered that we could ignore spectroscopic interference of oxidative molybdenum in determining the ^{113}Cd and ^{114}Cd contents in the ICP-MS analysis.

As shown in Fig. 3, after desorption of apoplastic ^{113}Cd by excessive ^{114}Cd, distribution of ^{113}Cd and ^{114}Cd in roots is as follow. (1) apoplastic bound ^{114}Cd is derived from desorption solution of excessive ^{114}Cd. (2) apoplastic bound ^{113}Cd is derived from desorption solution of excessive ^{114}Cd. (3) symplastic ^{113}Cd is derived from ^{113}Cd- uptake experiment. Therefore, ^{113}Cd content in roots is the sum of (1) and (2). Symplastic ^{113}Cd is the subtraction between total ^{113}Cd and ^{113}Cd derived from an enriched stable of ^{114}Cd. As shown in Fig. 1, the total ^{113}Cd contents in the roots signifies the ^{113}Cd contents in the roots after the desorption experiment (Fig. 1). The total [113]Cd content in the roots at 40 and 400 nmol Cd was 23.0 ± 4.3 and 87.7 ± 5.6 mg kg[-1] (dry weight), respectively (Table2). In contrast, the [114]Cd content at 40 and 400 nmol Cd was 117.3 ± 9.4 and 644.5 ± 33.7 mg kg[-1] (dry weight), respectively (Table2). The purification rate of the [114]Cd-enriched stable isotope used in the present study was 93.60%; whereas, the composition rate of [113]Cd in the [114]Cd-enriched stable isotope was 5.6%. The total [114]Cd content in the roots after desorption of 20 μmol [114]Cd was approximately 5.5-fold higher than that using 2 μmol [114]Cd (Table 2),suggesting that the apoplastically bound [113]Cd content, derived from the enriched isotope [114]Cd, increased with an increase in the concentration of [114]Cd in the desorption solution. Actually, the apoplastically bound [113]Cd contents, derived from the enriched isotope [114]Cd (2 and 20 μmol) were 6.6 ± 0.5 and 36.6 ± 1.8 mg kg[-1], respectively (Table 2); these values were calculated using equation in Fig.3. The contribution rate of [113]Cd content derived from the enriched stable isotope of [114]Cd for total [113]Cd in the roots was 28.6% for the 40 nmol [113]Cd treatment. In contrast, the contribution rate of [113]Cd content derived from [114]Cd for total [113]Cd content in the roots was 41.8% for the 400 nmol [113]Cd treatment (Table 2). These results indicate that the [113]Cd derived from the enriched stable isotope of [114]Cd must be subtracted from the total [113]Cd content in the roots to evaluate the symplastic [113]Cd in the roots. The symplastic [113]Cd contents for the 40 and 400 nmol treatments, calculated using equation in Fig.3, were 16.4 ± 3.7 and 51.0 ± 3.8 mg kg[-1], respectively (Table 2). In the present study, we disregarded the contribution of [114]Cd derived from the enriched isotope of [113]Cd because the composition rate of [114]Cd in the enriched isotope of [113]Cd was considerably lower than that of [113]Cd in the enriched isotope of [114]Cd.

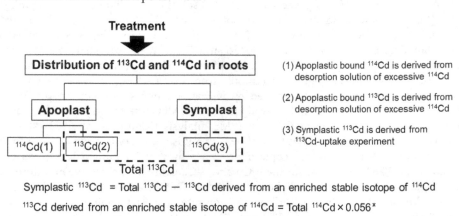

Fig. 3. Calculation of symplastic [113]Cd content in roots.

4.3 Comparison of the symplastic Cd contents in the roots between the two methods

To examine the validity of the new method for evaluating the symplastic Cd content in roots using [113]Cd and [114]Cd enriched isotopes, we compared the symplastic Cd content in roots using differences in the amounts of Cd absorbed at 2°C and 25°C with unlabeled Cd with the results obtained in the present study using the new method. In conventional Cd-absorption experiments, the Cd contents in roots at 40 and 400 nmol Cd in a 25°C treatment were 19.2 ± 1.6 and 84.4 ± 3.4 mg kg[-1] (dry weight), respectively (Table 3). In contrast, the Cd contents in roots at 40 and 400 nmol in the 2°C treatment were 4.1 ± 0.3 and 28.1 ± 0.73 mg kg^{-1} (dry weight), respectively.

The symplastic Cd contents at 40 and 400 nmol were estimated to be 15.1 ± 1.3 and 56.4 ± 2.7 mg kg^{-1}, respectively, which was evaluated using the difference in the amount of Cd absorbed at 2°C and at 25°C.

In the ^{113}Cd-absorption experiment, the symplastic ^{113}Cd contents in the roots at the 40 and 400 nmol ^{113}Cd treatments were 16.4 ± 3.7 and 51.0 ± 3.8 mg kg^{-1}, respectively (Table 2, 3). Therefore, the symplastic ^{113}Cd content after using the enriched isotopes was similar to the symplastic Cd content evaluated from the difference between the amount of Cd absorbed at 2°C and at 25°C. These results indicate that it is possible to evaluate the contents of symplastic Cd in roots using ^{113}Cd and ^{114}Cd enriched isotopes using the method proposed in the present study.

There have been many reports on Cd absorption in roots eliminating apoplastic bound Cd in Durum wheat, soybean and hyperaccumulator plants, such as *Thlaspi caerulescens* (Cataldo et al. 1983; Hart et al. 1998, 2002, 2006; Zhao et al. 2002). In these studies, the symplastic Cd content in the roots was determined by subtracting the Cd content in the roots at 2°C from the Cd content in the roots at 25°C; the Cd content was determined using a radioisotope of ^{109}Cd or a metabolic inhibitor. These methods have frequently been used to evaluate nutrient element absorption in roots. Radioisotopes in solute were the most useful markers used in these studies because they are chemically similar to the solute and can be distinguished from non-labeled solutes already contained in the roots (Davenport 2007). However, there are limitations to this method, including radioisotope administrative restriction and the restricted halflife of the radioisotope. Although the method involving a temperature difference between 2 and 25°C that was used in the present study is easy to handle because there is no radioisotope administrative restriction, there is, however, a limitation to this method: the symplastic Cd content in the roots cannot be evaluated using the same seedlings. This method has the advantage of no radioisotope administrative restriction and no restrictive radioisotope half-lives. In addition, this method uses half the number of seedlings that are required for the method using the temperature difference between 2 and 25°C because the symplastically absorped Cd in the roots can be evaluated using roots from the same seedlings. In addition, the method proposed in the present study is applicable to other plants, not only *S. melongena*. We indicated that it is possible to evaluate symplastic Cd in roots using ^{113}Cd and ^{114}Cd enriched isotopes. The proposed method will contribute to research on symplastic ion absorption in plant roots stated below.

40nM			
Total ^{114}Cd	Total ^{113}Cd	^{113}Cd derived from enriched ^{114}Cd	Symplastic ^{113}Cd
117.3±9.3	23.0±4.3	6.6±0.53	16.4±3.7
400nM			
Total ^{114}Cd	Total ^{113}Cd	^{113}Cd derived from enriched ^{114}Cd	Symplastic ^{113}Cd
644.5±33.7	87.7±5.6	36.6±1.8	51.0±3.8

Table 2. ^{114}Cd and ^{113}Cd content in roots (modified from Mori et al. 2009a)

40nM			
Symplastic 113Cd	**Cd(25°C-2°C)**	**Cd(25°C)**	**Cd(2°C)**
16.4±3.7	15.1±1.3	19.2±1.6	4.1±0.3
400nM			
Symplastic 113Cd	**Cd(25°C-2°C)**	**Cd(25°C)**	**Cd(2°C)**
51.0±3.8	56.4±2.7	84.4±3.4	28.1±0.7

Table 3. Comparison of the symplastic Cd content in roots (modified from Mori et al. 2009a)

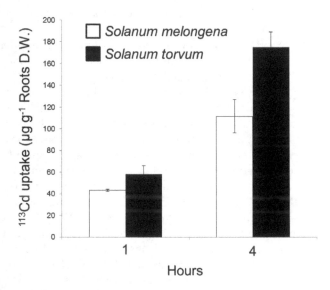

Fig. 4. Symplastic Cd absorption in roots of *Solanum melongena* and *Solanum torvum* with time. Experiment method is followed by the procedure illustrated in Fig.1 (modified from Mori et al.2009b)

We used the new method using enriched stable isotopes for evaluation of symplastic Cd absorption in roots of *solanaceous* plants (*Solanum melongena* and *Solanum torvum*) with contrasting root-to-shoot Cd translocation efficiencies (Mori et al. 2009a,b).

It is well known that efficiency of Cd translocation from roots to shoots is significantly higher in *S. melongena* than *S. torvum* (Arao et al. 2008, Mori et al. 2009a,b, Yamaguchi et al. 2011). Takeda et al.(2007) found that the Cd concentration in eggplant fruits could be reduced by grafting with *Solanum torvum* rootstock. Additionally, Arao et al.(2008) reported that although the Cd accumulation in shoots of *S. torvum* was lower than that found in *S. melongena*, there was no difference in the Cd content in roots of both plants when grown in culture solution. This result suggests that *S. torvum* develops noteworthy physiological mechanisms to suppress Cd translocation from roots to shoots, corresponding to the results observed in previous reports (Arao et al., 2008). Arao et al. (2008) suggested that symplastic Cd absorption and xylem loading capacity might be ascribed to the difference of Cd concentration in the shoots of *S. melongena* and *S. torvum*. We evaluated the symplastic Cd absorption rate in roots using enriched isotopes ^{113}Cd and ^{114}Cd. In time course-dependent experiments, the symplastic ^{113}Cd absorption rate for both plants increased with time (Fig. 4). In addition, the symplastic ^{113}Cd absorption rate of *S. melongena* was slightly higher than that of *S. torvum* at 4 h (Fig. 4). We examined kinetics analysis by similar method using enriched stable isotopes of ^{113}Cd and ^{114}Cd (Mori et al. 2009b). A kinetic study revealed that the symplastic Cd concentrations in the roots increased with increasing external Cd concentrations, but saturated at a higher concentration. The saturated curve obtained in this study suggests that absorption in both cultivars is mediated by a transporter that exhibits a similar affinity for Cd.. Moreover, the symplastic Cd concentrations slightly differed between the roots of *S. melongena* and *S. torvum*. Based on the reaction curves obtained, the K_m value was estimated to be 380 and 352 nmol L^{-1} for *S. melongena* and *S. torvum*, respectively. The corresponding V_{max} values were 152 and 101.5 μ g root dw^{-1} 0.5 h^{-1}. The V_{max} value of *S. melongena* was approximately 1.5-fold higher than that of *S. torvum*, which suggests that the density of the Cd transporter in the root cell membranes of *S. melongena* is higher than in *S. torvum*. In this experiments, If the symplastic Cd absorption in roots is estimated by the conventional method using the difference of temperature at 2 and 25°C, it is required time consuming and double seedlings for experiment preparation.

5. Conclusion

For biological system analysis, the application of ICP-MS in enriched stable isotope tracer experiments has increased because ICP-MS has now become the preferred technique. An enriched stable isotope technique would be potent and useful tool for biological system experiments including element uptake, distribution and chemical form in plants. In this chapter, we introduced our one example of element uptake system using enriched isotope of ^{113}Cd and ^{114}Cd. This method has several merits compared to conventional methods if ICP-MS instrument is able to use. Application of enriched isotopes such as ^{113}Cd and ^{114}Cd would attain a new insight for plant biological system and will become a new tool to evaluate element behavior in plants.

6. Acknowledgment

This work was partly supported by the Program for the Promotion of Basic Research Activities for Innovative Biosciences (PROBRAIN).

7. References

Arao, T.; Takeda, H. & Nishihara, E. (2008). Reduction of cadmium translocation from roots to shoots in eggplant (Solanum melongena) by grafting onto Solanum torvum rootstock. *Soil Science and Plant Nutrition*, Vol. 54, pp.555–559.

Cataldo, D.A.; Garland, T.R. & Wildung, R.E. (1983). Cadmium uptake kinetics in intact soybean plants. *Plant Physiology*, Vol.73, pp.844–848.

Codex Alimentarius Commission 2005: Joint FAO/WHO Food Standards Programme. Twenty-eighth session, 4–9 July 2005, Rome, Italy. Report of the 37th session of the Codex Committee on Food Additives and Contaminants, 25–29 April 2005. Para. 175, Appendix XXVI. The Hague, the Netherlands. Available at URL: http://www.codexalimentarius. net/web/reports.jsp]ALINORM 05/28/12

Dannel, F.; Pfeffer, H. & RÖmheld, V. (2000). Characterization of root boron pools, boron uptake and boron translocation in sunflower using the stable isotopes ^{10}B and ^{11}B. *Australian Journal of Plant Physiology*, Vol.27, pp.397-405.

Davenport, R.J. (2007). Ion uptake by plant roots. In Plant Solute Transport. Eds AR Yeo and TJ Flowers, pp. 193-213. Blackwell Publishing, Oxford.

Epstein, E. (1973). Mechanisms of ion transport through plant cell membranes. *International Review of Cytology*, Vol.34, pp.123-168.

Epstein, E. & Hagen, C.E. (1952). A kinetics study of the absorption of alkali cations by barley roots. *Plant Physiology*,Vol.27, pp.457-474.

Glass, A.D.M. (2007). The apoplast: a kinetic perspective. In The Apoplast of Higher Plants: Compartment of Storage,Transport, and Reactions. The Significance of the Apoplast for the Mineral Nutrition of Higher Plants. Eds B Sattelmacher and WJ Horst, pp. 87-96. Springer-Verlag, Dordrecht.

Hart, J.J.; Welch, R.M.; Norvell, W.A. & Kochian, L.V. (2002). Transport interactions between cadmium and zinc in roots of bread and durum wheat seedlings. *Physiologia Plantarum*, Vol.116, pp.73-78.

Hart, J.J.; Welch, R.M.; Norvell, W.A. & Kochian, L.V. (2006). Characterization of cadmium uptake, translocation and storage in near-isogenic lines of durum wheat that differ in grain cadmium concentration. *New Phytologist*, Vol.172, pp.261-271.

Hart, J.J.; Welch, R.M.; Norvell, W.A.; Sullivan, L.A. & Kochian, L.V. (1998). Characterization of cadmium binding, uptake, and translocation in intact seedlings of bread and durum wheat cultivars. *Plant Physiology*, Vol.116, pp.1413-1420.

Kawasaki, A. & Oda, H. (2005). Application of ^{113}Cd as a tracer in evaluation of cadmium uptake by soybean under field conditions. Japanese *Journal of Soil Science and Plant Nutrition*, Vol.76, pp.261-267 (in Japanese with English summary).

Kawasaki, A.; Oda, H. & Yamada, M. (2004). Application of enriched ^{113}Cd-tracer to the soil pot experiment of soybean plant. *Japanese Journal of Soil Science and Plant Nutrition*, Vol.75, pp.667-672 (in Japanese with English summary).

Kimura, K.; Yoshida, K.; Sugito, T. & Yamasaki, S. (2003). Correction of interference by Mo oxide on Cd measurement by ICPMS. *Japanese Journal of Soil Science Plant Nutrition*, Vol.74, pp.493-497 (in Japanese with English summary).

Keller, C. (1995). Application of centrifuging to heavy metal studies in soil solutions. *Communication Soil Science and Plant Analysis*, Vol.26, pp.1621-1636.

Lasat, M.M.; Baker,A.J.M. & Kochian L.V. (1996). Physiological Characterization of Root Zn^{2+} Absorption and Translocation to Shoots in Zn Hyperaccumulator and Nonaccumulator Species of *Thlaspi. Plant Physiology*, Vol.112, pp.1715-1722.

Lasat, M.M.; Baker, A.J.M. & Kochian, L.V. (1998). Altered Zn compartmentation in the root symplasm and stimulated Zn absorption into the leaf as mechanisms involved in Zn hyperaccumulation in *Thlaspi caerulescens. Plant Physiology*, Vol.118, pp.875-883.

Ma, J.F.; Mitani, N.; Nagao, S.; Konishi, S.; Tamai, K.; Iwashita, T. & Yano, M. (2004). Characterization of the Silicon Uptake System and Molecular Mapping of the Silicon Transporter Gene in Rice. *Plant Physiology*, Vol. 136, pp. 3284-3289

May, T.W. & Wiedmeryer, R.H. (1998). A table of polyatomic interferences in ICP-MS. *Atomic Spectroscopy*, Vol.19, pp.150-154.

Ministry of Agriculture Forestry and Fisheries of Japan 2002:Survey of the cadmium contained in domestic vegetables. Available at URL:http://www.maff.go.jp/cd/PDF/C12. pdf,p34, p38, p25 (in Japanese)

Mitani, N. & Ma, J.F. (2005). Uptake system of silicon in different plant species. *Journal of Experimental Botany*, Vol. 56, pp. 1255–1261.

Mori, S.; Kawasaki, A.; Ishikawa, S. & Arao, T. (2009a). A new method for evaluating symplastic cadmium absorption in the roots of Solanum melongena using enriched isotopes [113]Cd and [114]Cd. *Soil Science and Plant Nutrition*, Vol.55, pp. 294–299.

Mori, S.; Uraguchi, S.; Ishikawa, S. & Arao, T. (2009b). Xylem loading process is a critical factor for determining Cd accumulation in the shoots of Solanum melongena and Solanum torvum. *Environmental and Experimental Botany*, Vol.67, pp.127-132.

Nakanishi, H.; Ogawa, I.; Ishimaru, Y.; Mori, S. & Nishizawa, K.N. (2006). Iron deficiency enhances cadmium uptake and translocation mediated by the Fe^{2+} transporters OsIRT1 and OsIRT2 in rice. *Soil Science and Plant Nutrition*, Vol.52, pp.464–469

Oda, H.; Yada, S.; Kawasaki, A. (2004). Uptake and transport of Cd supplied at different growth stages in hydroponically cultured soybean plants. *Biomedical Research on Trace Element*, Vol.15, pp.289–291 (in Japanese with English summary).

Rains, D.W.; Epstein, E.; Zasoski, R.J. & Aslam, M. (2006). Active silicon uptake by wheat. *Plant and Soil*, Vol.280, pp.223–228.

Stürup, S.; Hansen, H.R. & Gammelgaard, B. (2008). Application of enriched stable isotopes as tracers in biological systems: a critical review. *Analytical and Bioanalytical and Chemistry*, Vol.390, pp.541–554.

Takano, J.; Noguchi, K.; Yasumori, M.; Kabayashi, M.; Gajdos, Z.; Miwa, K.; Hayashi, H.; Yoneyama, T. & Fujiwara, T. (2002). Arabidopsis boron transporter for xylem loading. *Nature*, Vol.420, pp.337-340.

Takeda, H.; Sato, A.; Nishihara, E. & Arao, T. (2007). Reduction of cadmium concentration in eggplant (*Solanum melongena*) fruits by grafting with solanum torvum rootstock. *Japanese Journal of Soil Science and Plant Nutrition*, Vol.78, pp.581–586 (in Japanese with English summary).

Ueno, D.; Iwashita, T.; Zao, F.J. & Ma, J.F. (2008). Characterization of Cd translocation and identification of the Cd form in xylem sap of the Cd-hyperaccumulator *Arabidopsis halleri*. *Plant and Cell Physiology*, Vol.49, pp.540–548.

Ueno, D.; Koyama, E.; Kono, I.; Ando, T.; Yano, M. & Ma, J.F. (2009). Identification of a Novel Major Quantitative Trait Locus Controlling Distribution of Cd Between Roots and Shoots in Rice. *Plant and Cell Physiology*, Vol. 50, pp.2223-2233

Ueno, D.; Ma, J.F.; Iwashita, T., Zhao, F.J. & McGrath, S.P. (2005). Identification of the form of Cd in the leaves of a superior Cd-accumulating ecotype of *Thlaspi caerulescens* using [113]Cd-NMR. *Planta*, Vol.221, pp.928–936.

Uraguchi, S.; Mori, S.; Kuramata, M.; Kawasaki, A.; Arao, T. & Ishikawa, S. (2009). Root-to-shoot Cd translocation via the xylem is the major process determining shoot and grain cadmium accumulation in rice. *Journal of Experimental Botany*, Vol.60, pp.2677-2688

Wanger, G.J. (1993). Accumulation of cadmium in crop plants and its consequences to human health. *Advanced Agronomy*, Vol.51, pp.173–212.

Yada, S.; Oda, H. & Kawasaki, A. (2004). Uptake and transport of Cd supplied at early growth stage in hydroponically cultured soybean plants. *Biomedical Research on Trace Elements*, Vol.15, pp.292–294 (in Japanese with English summary).

Yamaguchi, N.; Mori, S.; Baba, K.; Yada, S.; Arao, T., Kitajima, N.; Hokura, A. & Terada, Y. (2011). Cadmium distribution in the root tissues of *solanaceous* plants with contrasting root-to-shoot Cd translocation efficiencies. *Environmental and Experimental Botany*, Vol.71, pp.198–206

Yu, Q.; Tang, C.; Chen, Z. & Kuo, J. (1999). Extraction of apoplastic sap from plant roots by centrifugation. *New phytologist*, Vol.143, pp.299-304.

Zhao, F.J.; Hamon, R.E.; Lombi, E.; McLaughlin, M.J. & McGrath, S.P. (2002). Characteristics of cadmium uptake in two contrasting ecotypes of the hyperaccumulator *Thlaspi caerulescens. Journal of Experimental Botany*, Vol.53, pp.535–543.

7

Determination of Actinides Using Digital Pulse Processing Analysis

B. de Celis[1], V. del Canto[1], R. de la Fuente[1],
J.M. Lumbreras[1], J. Mundo[2], and B. de Celis Alonso[3]

[1]*Escuela de Ingenieria Industrial, Universidad de Leon*
[2]*Universidad Autonoma de Puebla*
[3]*Universidad de Erlangen,*
[1]*Spain*
[2]*Mexico*
[3]*Germany*

1. Introduction

The determination of low concentrations of actinides (Pu, U, Am, Np, etc) in environmental samples is vital for evaluating radioactive contamination caused by nuclear reactors, atomic bomb tests, and any type of nuclear incident. In an emergency situation, rapid analytical methods are essential to provide timely information to the authorities working to protect the environment and the population from the consequences of possible contamination.

Actinides disintegrate by emission of alpha particles which are difficult to detect because they are absorbed by the samples themselves. They are very dangerous to man because of the high relative biological effectiveness of alpha particles and their tendency to accumulate in several parts of the body (bones, kidney, liver, etc.) for many years because of their long biological and physical half-lives. Their determination usually demands lengthy analytical procedures which usually employ radiochemical pre-treatment of the sample followed by measurements using alpha spectrometry.

Alpha-emitting radioisotopes produce alpha particles at characteristic energies between 4 and 7 MeV, which can differ by as little as 10 keV, close to the smallest resolution of the silicon detectors used in alpha spectrometers. Alpha particles are heavy charged particles. Therefore, any physical medium between the alpha-emitting radionuclide and the detector will strongly absorb most of the alpha particle energy. These attenuations are produced by the sample itself or by any material between the sample and the detector. To solve this problem two different analytical procedures are commonly applied. A separation and concentration step is necessary in both procedures to avoid possible energy interferences between radionuclides and to concentrate in a small volume a significant amount of radioactivity, making it easier to measure. In order to account for any loss of the sample during separation, a known quantity of a specific isotope or tracer is usually added. The tracer is an isotope of the element under study with similar chemical behaviour. The radionuclide and tracer behaviour during the chemical treatment is the same, assuming that the tracer is homogeneously mixed and brought into chemical equilibrium with the sample.

The first procedure applies radiochemical sample treatment in an aqueous medium to separate and concentrate the different actinides before their measurement using high resolution passivated implanted planar silicon (PIPS) detectors (Holm, 1984; Aggarwal et al., 1985). This procedure allows for the alpha energies of the nuclides present in the sample to be identified with almost no interference. Thus very low minimum detectable activities (MDA) can be reached, but this process does require long analysis times. In this work we summarize this procedure and present results obtained in our laboratory using PIPS technology.

The second procedure using liquid scintillation (LS) is accurate and reproducible and considerably faster than other methods (Basson and Steyn, 1954; McKliven and McDowell, 1984; McDowell, 1986; Abuzwrda et al., 1987; Miglio and Willis, 1988; Lucas, 1980). The radionuclides are in a solution of an organic scintillator and there is no risk of sample self-absorption, giving a counting efficiency of nearly 100%. Recent improvements have been made using organic cocktails which act simultaneously as extractors and scintillators, enabling elements to be separated in the relatively short time of several hours. However, the identification of the different radioisotopes is not simple because the resolution is low.

The latest developments have focused on obtaining higher energy resolution by optimizing the chemical separation techniques and the extracting cocktails to provide the maximum amount of light emission and reduce quenching (McDowell, 1986). The electronics associated with these techniques have also been improved to reduce electronic noise and discriminate between beta and alpha pulses (McDowell, 1986; Reboli et al., 2005). The technique called photon-electron rejecting alpha liquid scintillation extracts the actinides from an aqueous solution into an organic phase containing the extractant, an energy transfer reagent, and a light emitting fluor. The counting efficiency reached is nearly 100% and the energy resolution is 5% for 4 to 6 MeV alpha particles. Using analogue pulse shape discrimination (PSD), it is possible to reduce the background from photo-electrons produced by external gamma-rays and to eliminate interference from beta emitters. PSD electronically selects pulses produced by alpha particles based on the longer decay time of their light emission (30 nsec), which is due to their much higher linear energy transfer to the scintillation solution.

The LS technique offers the advantage of a short analysis time but the resolution and the possibility of reducing interferences is limited. The initial way of increasing energy resolution was to use scintillation cocktails that produced more photons and improved light collection by using mirrors and a liquid between the sample and the photocathode to reduce light reflection (Hanschke, 1972; McDowell, 1994). The use of solid state detectors has also been used to improve resolution. Avalanche photodiodes (APD) have a larger gain than common photodiodes and are more sensitive to the frequency of the scintillator due to their greater spectral response. The replacement of photomultiplier tubes (PMTs) by APDs was successful for gamma-ray scintillation detection with energy resolution improvements and more compact detectors. Large area avalanche photodiodes have been applied to alpha liquid scintillation spectroscopy (Reboli et al., 2005). The influence of several parameters on energy resolution was studied: temperature, bias voltage, nature of the scintillating cocktail, and geometry of the counting vial. The improvement in energy resolution was attributed to the higher quantum efficiency and the more uniform active area of the photocathode compared to PMTs. The silicon photodiodes have a larger spectral response than PMTs, making it easier to match the fluorescence spectra of the scintillators with the APD spectral response. However, APDs have the typical limitations of photodiodes: a limited gain and a higher noise contribution.

In this review we describe the experimental work performed in our laboratory in applying digital cards to the analysis of actinides in environmental samples and we compare our methodology with the classical procedures described above. The first section shows some results obtained using PIPS detectors. The use of PSD with fast digital cards was applied to LS spectrometry. Finally, we studied the results obtained by applying coincident liquid scintillation/high resolution gamma spectrometry to reduce interference and increase the limits of detection.

2. High resolution alpha spectroscopy with passivated implanted planar silicon detectors

This procedure consists of the radiochemical separation of the radioisotopes of interest and their electro-deposition to form a few atomic layers in order to avoid the loss of alpha particle energy. Such attenuation can be produced by the sample itself or by any material between the sample and the detector. The result is a characteristic tail in the alpha peak (Fig.1). To reduce the size of the tail the samples are counted in a vacuum camera and the samples have to be as thin as possible to avoid self absorption. Electro-deposition, evaporation and precipitation are the most commonly used procedures to produce the thinnest samples possible. The sample is placed in front of the detector inside a vacuum camera and data are acquired for a preset period of time. Achieving the desired lower limit of detection requires very long times because of the low activities involved. Count times of one or two days are common, and so the total analysis time including sample preparation can be several days.

The proper preparation of the sample is an important step in achieving high quality results in alpha spectroscopy. Different methodologies have been published for specific applications (Glover, 1984; Greeman et al., 1990; Gomez et al., 1998; Sarin et al., 1990). There are three principal steps in the preparation of an alpha spectroscopy sample: preliminary treatment of the sample, chemical separation and preparation of the source to be measured.

The preliminary treatment is performed to homogenize the sample and to prepare it for the chemical processing. Different procedures are used for solid samples (e.g., food, soils, plants), liquid samples (water, blood, urine, etc.), and absorbents of gaseous or liquid samples (air filters, wipe type samples, etc.). In this early step the tracer nuclides are added to the samples.

Chemical separation is used to isolate and concentrate the elements of interest. Techniques used for separation include co-precipitation, solvent extraction, ion exchange and extraction chromatography.

The last step is required to prepare a sample suitable for being measured. The source preparation is an important step in obtaining the maximum resolution with an alpha spectrometer. The techniques used must be able to produce a very thin and uniform deposit to minimize the energy attenuation of alpha particles. The sources must be stable, and free of liquid, solvent and acid residues that could damage the vacuum camera or the detector. There are three main procedures for source preparation: electro-deposition (Singh et al.,1979), in which case the source is supposed to have a few atomic layers of the alpha emitter and the attenuation is extremely small; evaporation from an aqueous or organic solvent (Talvmtie, 1972; Aggarwal et al., 1985), which is sometimes followed by flaming under vacuum (Glover, 1984) to remove the organic material; and co-precipitation (Singh et al., 1979 and 1983).

Fig.1 shows an alpha spectrum of a solution containing Ra-226 taken with a Canberra A450-18AM PIPS detector, with 450mm^2 of active area, 18 keV of alpha resolution, and a background of 6 counts/day. The sample was obtained after the evaporation of a few microliter droplets of the radioactive solution on a metal disk. Despite the small amount of residue left, about 0.1 mg/cm^2, the attenuation was still significant as can be seen by the long tails of the alpha peaks. The peak energies of the alpha radionuclides could be easily determined by looking at the peak edge position but the large tails make their quantitative determination difficult.

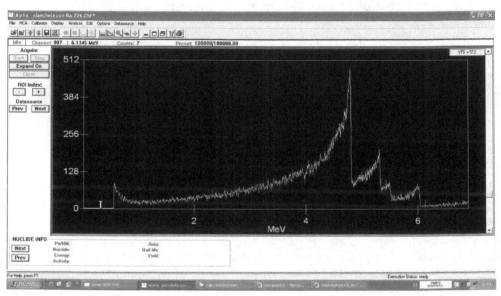

Fig. 1. Alpha spectrum of an evaporated solution of Ra-226 taken with a passivated implanted planar silicon detector. The software employed was Genie-2000, Canberra. The first peak situated to the left of the spectrum corresponds to Ra-226 (4.7 MeV), while subsequent peaks correspond to Po-210 (5.3MeV), Rn-222 (5.49MeV), and Pb-218 (6.0MeV).

Fig.2 shows the spectrum of a mineral sample containing U-238 and U-234. The sample was prepared by evaporation of the organic cocktail on a metal plate. The peaks of U-238 (4.2 MeV) and U-234 (4.7 MeV) are clearly visible but not the U-235 peak (4.5 MeV) that lies between both of them. The energy resolution was worse than in the case of aqueous evaporation because there was more organic residue (1 mg/cm^2). The size of both peaks was the same because the sample contained natural uranium, and the U-238 and U-234 were approximately in radioactive equilibrium. The energy of the alpha peaks could be determined but a quantitative analysis was not possible because of the large peak tails. Chemicals to spread the sample solution can be added prior to its evaporation to obtain more uniform deposition, and the organic deposits burned off before counting, but this causes poor adherence of the sample to the backing and poor resolution. In some cases it is possible to obtain a nearly solid-free deposit source sample. The procedure consists of evaporating the organic or aqueous sample to dryness, then treating with perchloric and nitric acids to oxidize the residual organic matter. The purified ion in aqueous solution is then extracted into thenoyltrifluoroacetone (TTA), deposited onto a stainless steel disc, and evaporated.

Sources are usually prepared by depositing micrograms of the element onto a flat, polished metal disk by electroplating from an aqueous medium. Properly prepared sources weigh less than 50 µg/cm². In this case a resolution of 50-100 keV for 4-7 MeV alpha-particles can be reached. Fig.3 shows the alpha spectrum of a soil sample containing U-238, U-234 , U-235, and U-232 (the yield tracer). A small peak corresponding to U-235 can also be observed between the U-238 and U-234 peaks. The energy resolution of the spectrum enables all the uranium isotopes to be identified. Quantitative analysis can also be performed because the peaks are clearly defined and their area can be easily evaluated. The background is very low and it is possible to reach very small MDAs.

Cooprecipitation offers an alternative to electrodeposition which is too time consuming. The actinides can co-precipitate with small amounts of a rare-earth element carrier. The carrier elements may be precipitated as a fluoride by addition of hydrofluoric acid. The precipitate is filtered onto a 0.1 µm membrane filter which is dried, mounted onto a support backing, and used for alpha spectroscopy. The method is fast, inexpensive, and produces resolution nearly as good as electrodeposition.

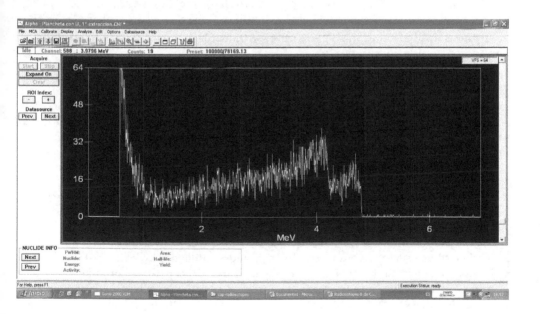

Fig. 2. Alpha spectrum of an evaporated organic extractant containing uranium. The U-238 (4.2 MeV) and U-234 (4.7 MeV) peaks (in blue) are clearly visible but not the U-235 peak (4.5 MeV) that lies between the two of them.

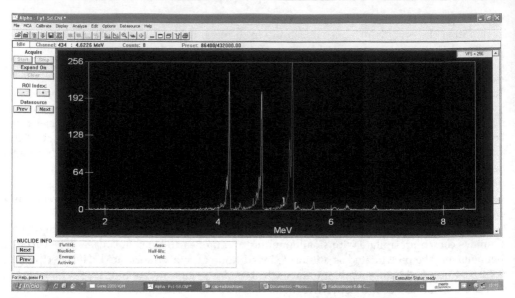

Fig. 3. Alpha spectrum of a soil sample containing U-238 (4.2 MeV), U-234 (4.7 MeV) (blue peaks), and U-232 (5.3 MeV) added as a yield tracer (red peak). A small peak corresponding to U-235 (4.4 MeV) can also be observed between the large peaks of U-238 and U-234.

3. Liquid scintillation alpha spectrometry using fast digital card and pulse shape discrimination analysis

The experimental equipment used in our laboratory to analyse actinides by LS spectrometry consisted of a standard photomultiplier tube with its pre-amplifier output directly connected to a fast digital card installed in a desktop computer. Two coincident photomultipliers were not needed, because the thermal electron noise in the phototubes was below the alpha pulse level. However, electronic noise, when superposed on the signal, could affect final resolution of the energy peaks. Usually, samples are immersed in a liquid to improve the transmission of light to the photomultiplier tube and they are surrounded by a light reflector to increase the efficiency of light collection. In our case the sample was placed directly above the photomultiplier tube. However, it has been observed that the geometry and reflector shape are important if good energy and pulse shape resolution are desired (McKlveen and McDowell, 1975).

The photomultiplier tube was a Saint Gobain RCA XP2412B with an AS07 preamplifier. It is possible to use several types of solid and liquid scintillators by placing them directly above the photomultiplier tube, but it is necessary to keep the phototube in a dark environment to avoid causing damage to the photocathode when a high voltage is applied. In our case the scintillator was the organic extraction cocktail containing the radionuclides of interest.

The typical analytical procedures to treat the solid samples were drying, grinding, and combustion of the sample to remove organic matter. Next, the sample was dissolved in a strong acid solution (nitric, perchloric and hydrofluoric acids). When samples contained silicate material, hydrofluoric acid was required, and repeated evaporation of the sample dissolved in acid was needed to remove the silicates as silicon tetrafluoride. For the general case of samples without significant amounts of thorium or iron a two-stage extraction is a fast, simple separation procedure (Cadieux et al., 1994; McDowell, 19869). A first extraction from 0.8M nitric acid separated U and Pu into the organic phase. After the organic phase was removed, the aqueous phase was adjusted to a pH of 2.5 to 3.0 by the addition of a formate buffer solution. The second extraction is then performed to remove americium and curium. The extractive scintillator mixture contains 120g/l of HDEHP, 180g/l of scintillation grade naphthalene and 4.0g/l of 2-(4'-biphenylyl-6-phenyl-benzoxazole) [PBBO] dissolved in spectroscopic grade p-xylene. After the phases were separated the oxygen was purged to reduce quenching and improve energy resolution. Tests of the two-step extraction on water samples with known actinide activities have shown recoveries higher than 95% with a precision of 2% to 3% (Cadieux et al. 1994). In our case we found recoveries of 95% for the first extraction and about 80% for the second one with a precision of 5%.

PSD analysis is an essential part of the procedure and can be performed with analog or digital electronics. The recent availability of fast digital cards offers the advantage of greater reliability and simplicity of use. This is achieved because it removes the need for the electronic equipment necessary when employing conventional analog procedures, i.e. multichannel analysers (MCAs), amplifiers, coincidence electronics, single-channel analysers, etc. The direct analog-to-digital conversion of the detector signal and the use of software techniques allows to simplify tasks usually performed by the analog amplifier, i.e. electronic noise reduction, the treatment of pile-up pulses, trigger level restoration, and the reduction of ballistic deficit. MCAs are not needed because the digitized signal has all the necessary information to determine the pulse amplitude, which is proportional to the particle energy, and to obtain the energy spectrum. The direct analog-to-digital conversion of the detector signal offers many other possibilities (White and Miller, 1999; Warburton el al., 2000), such as alpha/beta/gamma discrimination by analysis of the electronic pulse shape (PSD), precise detection time determination of pulses by using time stamping information, the implementation of coincidence techniques, and dead-time reduction.

In order to test the accuracy of the procedure a standard liquid containing known activities of uranium (U-238, U-234, U-235), Am-241, and Sr-90 was employed for the analysis. The results from the first extraction, containing U-238, U-234, and U-235, and from the second extraction, containing Am-241, were place in two glass vials to be measured. The vials were situated directly in front of the photomultiplier tube to allow the maximum entrance of the light emitted by the sample. Other arrangements, adding reflectors to the vial or immersing it in liquid to improve light transmission, were also studied, but it was observed that different light paths produced a broadening of the energy peak and these procedures were abandoned.

The electronic signal at the pre-amplifier output was connected to a digital card (AlazarTechs ATS330 PCI) with the following technical specifications: two channels independently sampled at 12-bit resolution and a sampling rate of 50 million samples per second [Ms/sec], with multiple possibilities for triggering, multiple records and time stamping. The two channels are useful if coincidence measurements are performed with two independent electronic circuits and detectors. The device had been previously tested with different types of scintillation cocktails containing radioactive sources of known radioactivity and different emitting radionuclides (Ra-226, Sr-90, Am-141), to establish their pulse shape characteristics, the optimum bias voltage, and the card parameters (sampling rate, samples per pulse, triggering thresholds, voltage range, input impedance, and noise level). Those hardware conditions determined the optimum parameters for particle detection when the organic liquid scintillator was coupled with the photomultiplier tube.

The digital card was supplied with basic software to set up the acquisition hardware. It contained a software development kit to allow full control of the card. In order to develop a complete system able to perform alpha/beta/gamma spectrometry and coincidence experiments, many other functions were developed and incorporated in the original software. The system developed acts like a digital oscilloscope that visualizes each individual pulse, and also like a conventional MCA. Several selection functions were incorporated, such as voltage amplitude, record length, sampling rate, trigger level, input impedance. The characteristics incorporated into the MCA were identification of a region of interest, peak area determination, calibration of energies, dead-time determination, and loading and recording of spectra.

Several algorithms were also developed to perform PSD analysis. They separated the signals that came from different scintillator types or were caused by different type of particles, allowing simultaneous acquisition of alpha/beta/gamma spectra from a radioactive sample. In addition, AND and OR relations could be used during acquisition to obtain a compound spectrum from the sample.

Fig. 4 shows the spectrum of a soil sample containing uranium. The spectrum was obtained with the organic cocktail from the first chemical extraction. It was difficult to differentiate the peaks for U-238 (4.2 MeV) and U-234 (4.7 MeV), because the resolution is 10% (0.5 MeV), larger than the resolution (5%) reported by other researchers using specific instrumentation (McKlveen and McDowell, 1984; Cadieux, 1990) or employing avalanche photodiodes (Reboli et al., 2005). Different procedures were used to reduce electronic noise (i.e. elimination of noise harmonics by using fast Fourier transform), but the best energy resolution reached was 8%.

Fig. 5 shows the spectrum of a sample containing Ra-226. The peaks for Ra-226 (4.7 MeV) and its descendants, Rn-222 (5.49 MeV), Pb-218 (6.0 MeV), and Po-214 (7.69 MeV) interfered and formed a unique peak. The small peak to the right corresponded to Po-214 (7.69 MeV). The sample was immersed in an aqueous medium containing Packard alpha/beta Ultima-Gold scintillator and a few drops of Ultima-Gold F scintillator to improve the energy resolution which was 8%.

The counting efficiency was 80%. Efficiencies near 100% could probably be reached with better energy resolution and light collection. A disperse background of counts was observed to the left of the peaks. That was attributed to pulses of lower amplitude produced by more attenuated light paths. The background was measured with a sample of the pure cocktail extractor. Its value was 10^{-4} counts per minute [cpm] below the energy peak area. It could provide an MDA close to 10 Becquerels per kilogram [Bq/kg] in the case of a ten-gram sample and a measurement time of one day, assuming the absence of energy interference with other radionuclides. In any case, the results obtained in terms of energy resolution and MDA are very similar to those obtained by using a Packard TRICARB LSC 2900TR.

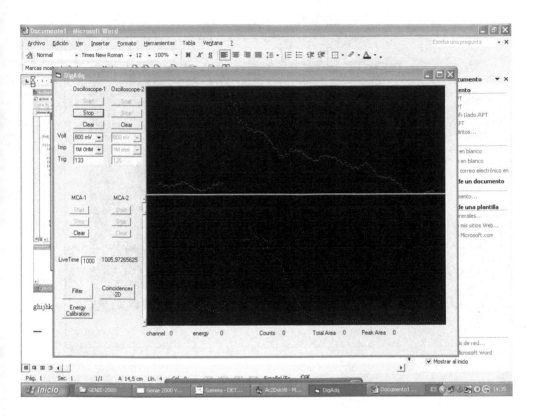

Fig. 4. Alpha spectrum of a uranium sample obtained by liquid-liquid extraction with organic solvents. The energy resolution was about 10% (0.5 MeV) and the peaks of U-238 (4.2 MeV) and U-234 (4.7 MeV) interfered. The upper part of the figure shows the shape of the electronic pulse at the preamplifier output. Each point of the pulse graph corresponds to 20 nsec.

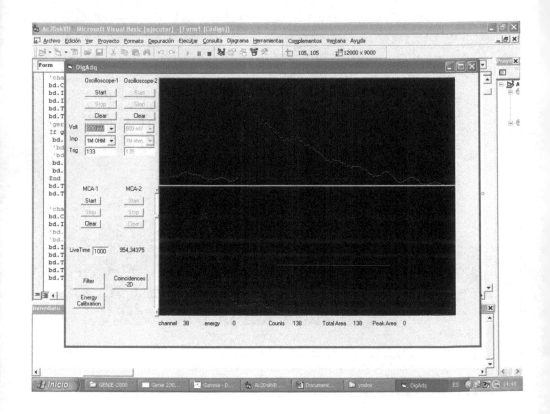

Fig. 5. Alpha spectrum of a radium sample in an aqueous medium using alpha/beta Ultima-Gold scintillator. The large peak corresponds to Ra-226 (4.7 MeV) and its descendents Po-210 (5.3 MeV), Rn-222 (5.49 MeV), and Pb-218 (6.0 MeV). They interfered and formed a unique peak. The small peak to the right is Po-214 (7.69 MeV). The upper part of the figure shows the shape of the electronic pulse at the preamplifier output. Each point of the pulse graph corresponds to 20 nsec.

4. Coincidence alpha X-ray in actinides using fast digital cards and pulse shape discrimination analysis

In order to increase the sensitivity of the procedure and reduce energy interference, gamma pulses emitted by actinides were measured in coincidence with alpha pulses. Many actinides decay by alpha particle emission in coincidence with low energy gamma/X-rays. When the alpha energies of two radionuclides interfere because their energy separation is below the energy resolution limit of LS spectroscopy, the coincidence procedure allows to obtain additional information by measuring the energy of the gamma/X-ray radiation coincident with the alpha emission.

The coincidence procedure is easy to implement by using a two channel digital card. The electronic procedure consists of recording on the second channel the X or gamma radiation coinciding with the alpha particles recorded on the first one. Using this method, it is possible to obtain X and gamma spectra with a significant reduction in background and the typical lead shield is not needed (de Celis, de la Fuente et al., 2007; de la Fuente, de Celis et al., 2008).

The detector employed for gamma radiation was a low-energy germanium, Canberra GL1515 with a 0.5mm Be window and a FWHM resolution of 304 eV at 5.9 keV and 551 eV at 122 keV. The detector area was 1500 square millimetres and its thickness 15 mm, which guarantees a high gamma-collection efficiency and a low energy resolution.

Tests were conducted with an Am-241 liquid source of known activity dissolved in 1ml of liquid scintillator. Am-241 emits 5.4 MeV alpha particles in coincidence with 59 keV gamma-rays. A plastic vial containing the radioactive source was placed directly over the photomultiplier tube and the germanium detector was situated above the sample. Small vials about 1cm in height were fabricated to reduce the distance between the liquid sample and the germanium detector. To implement the procedure in the digital card, the alpha trigger of the first channel is set "on" and the gamma trigger of the second channel is set to "off", in order to record gamma/X ray radiation only when the alpha channel is active. The pulses from the two detectors were stored jointly with the detection times and coincidences are identified by comparing the arrival time of the pulses proceeding from the liquid detector (alpha particles) and germanium detector (gamma/X rays). A delay time between pulses smaller than 1µs was considered to indicate a coincidence event.

Experimental tests were performed with a soil sample of known activity, previously reduced to ash and dissolved in nitric acid. The sample contained Am-241 and Pu-238 with activities below 1 Bq/kg. In conventional analysis most of these radionuclides are only detectable if radiochemical separation is combined with alpha spectrometry using PIPS detectors. With this procedure it was possible to identify and determine quantitatively the Am-241 activity. The MDA of the procedure was 0.01Bq/kg, which is very low compared with the 10 Bq/kg recorded when using directly gamma spectrometry and lower than the 0.1 Bq/kg achieved when using PIPS detectors. This is due to the high background reduction of the coincidence procedure. Other alpha/X-ray radionuclides, such as Pu-238, were not identified because of the efficiency reduction in the 10-20 keV region by the Ge absorption edge and X-ray self-absorption in the liquid sample. Tests to increase the MDA at these energies could be carried out with Si(PIN) detectors or using a phosphor sandwich [phoswich] detector with a higher resolution scintillator. Fig. 6 shows a coincidence event (alpha/gamma) and the coincidence gamma spectrum of a soil sample. The gamma pulse was delayed with respect to the alpha pulse because an analog amplifier with a shaping constant of 1μs was employed to reduce electronic noise and obtain a better resolution in the germanium detector. Electronic noise reduction in the germanium chain could also be achieved using digital procedures.

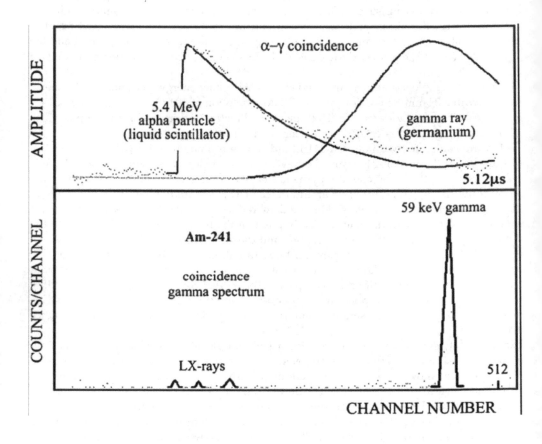

Fig. 6. Am-241 alpha/gamma coincident pulses (above) and coincident gamma spectrum (below) using liquid scintillation and a low energy germanium detector.

The procedure has the advantage of using the same geometry and chemical preparation for samples and standards, reducing the need for corrections for quenching. Interferences by beta/gamma coincidences are easily avoided, as beta particle pulses detected in the liquid scintillator are usually of smaller amplitude than alpha particle pulses. However, PSD, using the different scintillation decay time of alpha and beta particles, is intended for use in future. The use of a digital PSD system offers many other advantages, particularly the fact that the background of the detection system, which ultimately determines the sensitivity of the procedure, can be further reduced by studying the shape and detection time of the coincident particles, excluding those pulses which do not meet certain conditions.

5. Conclusions

Analytical procedures used to determine actinides in environmental samples were reviewed. The standard procedure using PIPS detectors is the most often used. It allows very small MDA to be reached due to the high resolution and very low backgrounds of these detectors. However, the analysis time is long (several days) and impossible to apply when many samples need to be processed or it is necessary to know the results in record time. LS spectrometry overcomes this inconvenience but at the expense of a poor resolution which makes it difficult to eliminate some interferences and thus determine quantitatively certain radionuclides. The LS analysis time is very short and improvements could be made to increase the light emission from the cocktail, reducing quenching and improving energy resolution. Future photodiode detectors could also help to improve energy resolution. A digital system is the natural way to implement many of these new developments due to its advantages in treating the digitized pulses. The MDA is ultimately dependent on the background of the technique. In our case it was possible to reduce the background by eliminating electronic signals which did not correspond with the correct pulse shape and also random electronic noise using standard digital procedures. Discrimination between different types of particles could be implemented by measuring the pulse light decay time without the need to resort to analogue electronic equipment.

Coincidence experiments could be easily set up using digital cards with two or more channels. The coincidence time could be determined with resolution times between 20 nsec and 1 nsec for cards of 50 Ms/sec to 1 Gs/sec sampling rates. Two-parameter studies using coincidence and 2-D diagrams can help to determine certain types of actinides. The technique is usually applied with analogue electronic equipment to determine radio-xenons resulting from nuclear subterranean tests but the setup presents more difficulties than when using digital equipment. In the case of actinide identification the technique could be applied to eliminate interferences between radioisotopes of similar alpha energies but different gamma emissions.

In some cases, for instance, the analysis of certain plutonium isotopes (Pu-239 and Pu-240), the alpha energies are similar and also equal the X-ray energy emissions. In this case conversion electron spectrometry with PIPs detectors using the same electro-deposited samples prepared for alpha spectrometry could help in obtaining a quantitative determination of both radioisotopes.

X-ray spectrometry could also be used to determine the total activity of the different radioisotopes of an element by using a high-resolution germanium detector in coincidence with LS spectrometry. The coincidence experiment is necessary to eliminate the background of the germanium detector and improve the MDA limit. Si (PIN) detectors, cooled by the Peltier effect, could offer a simple alternative if detectors of large area and small electronic noise are built in the future. Modern Si (PIN) detectors can reach energy resolutions of 125 eV at 5 keV with very low backgrounds but at the cost of a low counting efficiency because of their small detector area.

6. Acknowledgments

This research was supported by the Consejo de Seguridad Nuclear (CSN) of Spain, and the REM (Red de Estaciones de Muestreo) and PVRAIN (Plan de Vigilancia Radiológica Ambiental en la Central Nuclear de Sta. Mª Garoña) programmes.

7. References

Abuzwrda M., Abouzreba S., Almedhem B., Zolotarev Yu.A. and Komarov N.A. (1987). Evaluation of a photoelectron-rejecting alpha liquid-scintillation, *Radioanal. Nucl. Chem.*, 111 (1987) 11.

Aggarwal S.K., Chourasiya G., Duggal R.K., Singh C.P., Rawat A.S., and H.C. Jain. (1985). A comparative study of different methods of preparation of sources for alpha spectrometry of plutonium. *Nuclear Instruments and Methods in Physics Research, Section A*, Volume 238, Issues 2-3,1 August 1985, Pages 463-468.

Cadieux, J.R. (1990). Evaluation of a photoelectron-rejecting alpha liquid-scintillation (PERALS) spectrometer for the measurement of alpha-emitting radionuclides. *Nuclear Instruments and Methods in Physics Research Section A: Accelerators, Spectrometers, Detectors and Associated Equipment*, Volume 299, Issues 1-3, 20 December 1990, Pages 119-122.

Cadieux, J.R., Clark S., Fjeld R.A., Reboul S. and Sowder A. (1994). Measurement of actinides in environmental samples by photon-electron rejecting alpha liquid scintillation. *Nuclear Instruments and Methods in Physics Research Section A: Accelerators, Spectrometers, Detectors and Associated Equipment*, Volume 353, Issues 1-3, 30 December 1994, Pages 534-538.

de Celis, B., de la Fuente, R., Williart, A., de Celis Alonso, B., (2006). Coincidence measurements in a/b/g spectrometry with phoswich detectors using digital pulse shape discrimination analysis. In: *ISRP10* Symposium, Coimbra, Portugal, 17-22 September 2006.

de Celis, B., de la Fuente, R., Williart, A. and de Celis Alonso, B. (2007). Coincidence measurements in α/β/γ spectrometry with phoswich detectors using digital pulse shape discrimination analysis, *Nuclear Instruments and Methods in Physics Research Section A: Accelerators, Spectrometers, Detectors and Associated Equipment*, Volume 580, Issue 1, 21 September 2007, Pages 206-209. ISSN 0168-9002.

de la Fuente R., de Celis B., del Canto V., Lumbreras J.M., de Celis Alonso B., Martín-Martín A. and Gutierrez-Villanueva, J.L. (2008). Low level radioactivity measurements with phoswich detectors using coincident techniques and digital pulse processing analysis *Journal of Environmental Radioactivity*, Volume 99, Issue 10, October 2008, Pages 1553-1557

Glover, K. M. (1984). Alpha-particle spectrometry and its applications. *Int. J. Appl. Radiat. Isot.* 35, 239, 1984.

Gomez Escobar, V., Vera Tome, F., Lozano and J.C., Martin Sanchez, A., (1998). Extractive procedure for uranium determination in water samples by liquid scintillation counting, *Appl. Radiat. Isot.* 49, 875-883.

Greeman, D.J. and Rose, A.W. (1990). Form and behaviour of radium, uranium and thorium in central Pennsylvania soils derived from dolomite. *Geophys. Res. Lett.* 17, 833-836., 1990.

Hanschke, T. (1972). High-resolution alpha spectroscopy with liquid scintillators by optimisation of the geometry, Ph.D. Thesis, Hannover Technical University, 1972.

Holm E. , (1984). Review of Alpha-Particle Spectrometric Measurements of Actinides, *Int. J Appl . Radiat . Isot .* 35 285.

Lucas L.L . (1980). The Standardization of Alpha-Particle Sources, ASTM STP 698 (1980) 342.

McDowell W.J. and McDowell L.B., *Liquid Scintillation Alpha Spectrometry*, CRC Press, Boca Raton, FL, USA, 1994.

McDowell W.J., Nucl. Sci. Ser. on Radiochem. Tech. NAS-NS 3116 (1986).

McDowell, W.J. (1980). Alpha liquid scintillation counting: Past, present, and future *The International Journal of Applied Radiation and Isotopes*, Volume 31, Issue 1, January 1980, page 23.

McKliven, J.W. and McDowell, W.J. (1984). Liquid scintillation alpha spectrometry techniques, *Nuclear Instruments and Methods in Physics Research*, Volume 223, Issues 2-3, 15 June 1984, Pages 372-376.

Miglio J.J. and Willis L.C ., (1988). Simultaneous liquid scintillation determination of [239]Pu and[241]Am in tissue, *J. Radioanal. Nucl . Chem.* 12 3 (1988) 517.

Reboli, A., Aupiais, J. and Mialocq, J.C. (2005). Application of large area avalanche photodiodes for alpha liquid scintillation counting, *Nuclear Instruments and Methods in Physics Research* Section A: Accelerators, Spectrometers, Detectors and Associated Equipment, Volume 550, Issue 3, 21 September 2005, Pages 593-602.

Sarin, M.M., Krishnaswami, S., Somayajulu, B.L.K. and Moore, W.S. (1990). Chemistry of uranium, thorium and radium isotopes in the Ganga-Brahmaputra river system. *Geochim. Cosmochim. Acta* 5, 1990.

Singh N.P. , Ibraim S.H., Cohen N. and Wrenn M.E.,(1979). Determination of plutonium in sediments by solvent extraction and α-spectrometry, *Anal. Chem.* 50, 357 (1979) Pages 265-274.

Singh N.P. and Wrenn M.E. (1983). Determination of alpha-emitting uranium isotopes in soft tissues by solvent extraction and alpha-spectrometry, *Talanta*, Volume 30, Issue 4, April 1983, Pages 271-274.

Singh N.P., McDonald and Wrenn, E. (1988). Determinations of actinides in biological and environmental samples,*Science of The Total Environment*,Volume 70, March 1988, Pages 187-203.

Talvitie N.A . (1972). Electrodeposition of actinides for alpha spectrometric determination. *Anal .Chem.* 44 (1972) 280-283.

Warburton, W.K., Momayezi, M., Hubbard-Nelson, B., Skulski, W., (2000). Digital pulse processing: new possibilities in nuclear spectroscopy. *Applied Radiation and Isotopes* 53, 913-920. 2000.

White, T. and Miller, W. (1999). A triple crystal phoswich detector with digital pulse shape discrimination for alpha/beta/gamma spectroscopy, *Nuclear Instruments and Methods in Physics Research* Section A: Accelerators, Spectrometers, Detectors and Associated Equipment 422, 144-147.

8

History of Applications of Radioactive Sources in Analytical Instruments for Planetary Exploration

Thanasis E. Economou

Laboratory for Astrophysics and Space Research, Enrico Fermi Institute,
University of Chicago, Chicago,
USA

1. Introduction

Space age started with the launch of Sputnik-1 more than 50 years ago. Since then we have visited all the planets, some of them many times with very capable spacecrafts that are equipped with sophisticated payload that are returning significant information about the composition, physical condition of their surfaces and atmospheres and much more information. However, at the beginning of the space age, there were no techniques and instruments available to be used in space and had to be invented, designed, built and tested for the harsh environmental conditions in space. For the first analytical instruments in space, transuranium artificial radioisotopes produced in the national laboratories, proved to be very useful for applications in space. In early sixties Anthony Turkevich and his group at the University of Chicago applied a novel technique -- the Rutherford backscattering -- that is based on the interaction of the alpha particles with matter, to devise an instrument to obtain in-situ the chemical composition of the lunar surface [1]. In addition to measuring the backscattered alpha particles, the instrument also measures the proton energy spectra derived from the (α,p) reaction of the alpha particles with some light elements in the analyzed sample. Since the bombardment of a sample with a beam of alpha particles and x-rays from the same source also produces specific characteristic x-rays that results in additional compositional information, an additional x-ray detector was added to the ASI instrument to detect the produced x-rays. Based on the successful performance of this instrument on the lunar missions, more advanced and miniaturized versions complimented with an x-ray mode were developed and used in many NASA, ESA and Russian missions to several planetary bodies.

2. The alpha scattering instrument for the lunar missions

The technique of alpha backscattering for obtaining the chemical composition of planetary bodies was described for the first time and in detail by A. Turkevich, 1961[1], 1968[2]; Patterson et al., 1965[3]; Economou et al., 1970[4], 1973[5]. The compositional information is obtained from the energy spectra of scattered alpha particles and protons generated in (α,p) reactions of alpha particles with the matter in the analyzed sample. As it is shown in Fig. 1, a beam of monochromatic alpha particles with energy of 6.1 MeV from ^{242}Cm alpha source is bombarding the sample to be analyzed. The resulting scattered alpha particles and protons from (α,p) reactions are detected by solid state Si detectors situated as close to 180° as possible for maximum separation of individual elements. The energy of scattered alpha particles depends mostly on the scattered angle and the mass of the nucleus A, from which it is scattered. The sensor head of the Alpha Scattering Instrument (ASI) containing all the detectors, sources and the first stages of the electronic preamplifiers is deployed to the surface of the Moon and it is exposed to the harsh lunar environmental conditions, as it is shown in Fig. 2. The rest of the ASI electronics is placed inside the spacecraft compartment that is more controlled for temperature extremes.

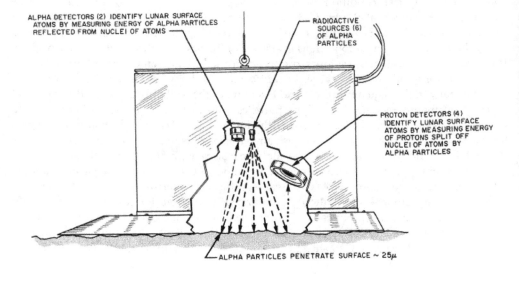

ALPHA DETECTORS (2) IDENTIFY LUNAR SURFACE ATOMS BY MEASURING ENERGY OF ALPHA PARTICLES REFLECTED FROM NUCLEI OF ATOMS

RADIOACTIVE SOURCES (6) OF ALPHA PARTICLES

PROTON DETECTORS (4) IDENTIFY LUNAR SURFACE ATOMS BY MEASURING ENERGY OF PROTONS SPLIT OFF NUCLEI OF ATOMS BY ALPHA PARTICLES

ALPHA PARTICLES PENETRATE SURFACE ~ 25μ

Fig. 1. The Alpha Scattering Instrument sensor head [2]

Fig. 2. The Alpha Scattering instrument deployed on the surface of the Moon

The energy E of scattered alpha particles in respect to its initial energy E_0 is a function of scattering angle θ and the mass A, of the target atom.

$$\frac{E}{E_0} = \left(\frac{4\cos\Theta + (A^2 - 16\sin^2\Theta)^{1/2}}{A+4} \right)^2$$

Which for $\theta = 180°$ converts to

$$\frac{E}{E_0} = \left(\frac{A-4}{A+4} \right)^2$$

Since we know the initial source energy, E_0, the mass A, from which an alpha particle is scattered, can be calculated from the measured energy E from the above equation. For the Surveyor lunar missions in the 1960's the ASI used several hundred millicuries of ^{242}Cm alpha particle source of 6.1 MeV. Figure 3 shows some alpha spectra from pure elements and the alpha and proton spectra from lunar samples from the Surveyor 7 lunar mission. From these spectra, the elemental composition of the lunar surface material has been derived. The ASI on three Surveyor series lunar missions in 1967-68 provided the first detailed and complete elemental composition of the lunar surface. The comparison of the lunar analyses results provided by the ASI on the Surveyor 5 and Surveyor 7 with those analyses of lunar material brought back from the Moon by the Apollo 11 astronauts is shown in Fig. 4 [6].

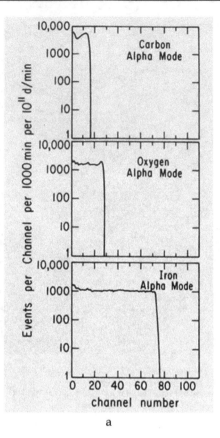

a

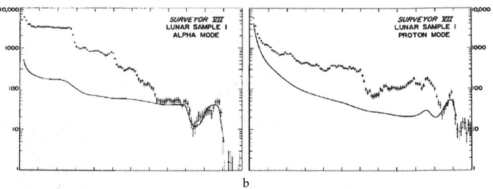

b

Fig. 3. a: Alpha spectra from carbon, oxygen and iron, b: ASI alpha and proton spectra from Lunar sample 1 of Surveyor 7.

This experiment was credited as the beginning of the Alpha Backscattering Spectroscopy (ABS) which has become a common analytical technique in many terrestrial laboratories. The ABS is using, however, a beam of alpha particles from a particle accelerator instead from radioisotopes.

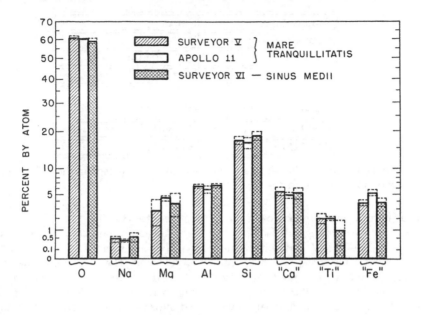

Fig. 4. Comparison of the ASI Surveyor 5 and Surveyor 7 results of the chemical composition of lunar surface material with those brought back by the Apollo 11 astronauts.

2.1 The ^{242}Cm alpha sources

The Alpha Scattering Instrument on the Surveyor lunar missions that provided the first chemical analysis of lunar material required alpha particle sources of high intensity and quality. In choosing the right isotope several criteria had to be taken into consideration. The decay half-live time has to be short in order to have essentially weightless sources, but long enough so that the sources will not decay appreciably before the completion of the measurements. Also, the source had to emit monoenergetic alpha particles with energy as high as possible and any decay daughter products should grow very slowly.

It was found that ^{242}Cm isotope ($T_{1/2}$ = 163 days, E_α=6.11 MeV) fits all the criteria best. This isotope emits essentially monoenergetic alpha particles and the decay half-life time of its daughter product ^{238}Pu is 86 yrs. So, it was ideal for the short missions to the Moon.

The procedure of preparing ^{242}Cm is long and complicated and it is described in details by Paterson et al., [7]. It starts with irradiation of ^{241}Am with neutrons in nuclear reactors which converts it to ^{242}Am by neutron capture followed by EC decay into ^{242}Cm and ^{242}Pu. After that, substantial purification yielded pure ^{242}Cm that was deposited on stainless steel plates of an active area of 2.8 mm in diameter and loaded into the ASI instrument.

Although ^{241}Am is wildly abundant because it is the decay product of reactor produced ^{241}Pu, the availability of ^{242}Cm is very scarce, especially in large quantities, due to its short half-life time. For the Surveyor lunar missions, a special program supported by the Atomic Energy Commission was undertaken at Argonne National Laboratory to produce adequate quantities of ^{242}Cm alpha sources with high qualities required for the experiment. Table 1 lists the main source characteristics of ^{242}Cm alpha sources used in the ASI on the Surveyor lunar missions in the late 1960's.

Radioisotope	^{242}Cm
Number of Sources	6
Total Intensity	120-470 millicuries
E_α	6.11 MeV
$T_{1/2}$	162.8 days
FWHM	1.5-1.9 %
FW0.1M	2.5-3.0%
FW0.01M	5.0-8.0%

Table 1. Characteristics of the ^{242}Cm alpha radioactive source used in the ASI instrument on the Surveyor 5-7 missions to the Moon.

2.2 The "Mini-Alpha" alpha proton X-ray spectrometer (APXS)

Most of the alpha emitting radioisotopes, beside emitting alpha particles, also emit x-rays in the energy range 14-20 keV and some gamma rays that are useful in exciting with different efficiency the characteristic XRF lines from every element. The higher Z elements are more efficiently excited by the X-ray lines emitted by the Curium isotopes alpha sources, while the lower Z elements are better excited by the alpha particles of 5-6 MeV (see Fig. 5a). These generated x-rays contain additional compositional information and were used to dramatically improve the analytical information of the ASI instrument, extending the performance to the presence of minor elements down to several tens of ppm range. On the other hand, the alpha mode is capable of better separating the light elements, while the x-ray mode has a better separation resolution for the heavier Z elements (see Fig. 5b). Therefore, an additional mode has been added to the original ASI instrument to detect the resulting characteristic x-rays that are produced when a sample is bombarded with a beam of alpha particles and x-rays from the same radioactive source. It is the combination of all three modes, the Rutherford alpha backscattering, the x-ray fluorescence (XRF) and the particle PIXE techniques that resulted in one integrated low power, low volume, but very powerful Alpha Proton X-ray Spectrometer (APXS) analytical instrument [8] that has been used in so many space missions [9-20].

A refined and miniaturized instrument with alpha, proton and x-ray modes (the "Alpha-Proton X-ray Spectrometer by Economou et al., 1976[8]) was proposed for analysis of Martian surface material during preparation for the Viking missions, but it was not selected for that mission. But the development of the "Mini-Alpha" APXS instrument was essential for future missions to Mars and other planetary bodies.

Figure 6 shows the diagram of the "Mini-Alpha" instrument. Its sensor head contains two telescopes with a combination of a thin dE/dx and thicker Si solid state detectors for the alpha and proton modes and a separate x-ray detector for the x-ray mode. The big challenge of miniaturization and enabling the use of such an instrument for space missions was the requirement of a good energy resolution for the x-ray detector. In the terrestrial laboratories every x-ray detector is kept at very low temperatures by cooling it with liquid nitrogen, which is for space application prohibitive. An enormous effort was undertaken to replace the Si-cooled x-ray detector with one that can operate at ambient Mars environment. Varieties of detectors made from exotic materials like Ge, HgI_2, CdTe, CdZn and other materials were tried with a different degree of success. Finally, we used a Si PIN detector that could provide sufficient energy resolution at martian ambient temperatures, and space qualify it at the last moment for the Pathfinder missions to Mars in 1996.

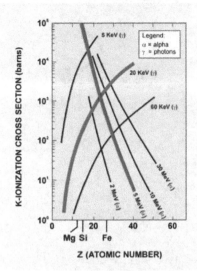

Fig. 5a. The characteristic X-rays are the result of two different mechanisms: X-ray fluorescence (in this particular case by the Pu L- lines in the 14-20 keV range) which is most effective for high Z elements and by particle induced X-ray emission (PIXE) that is most effective for the low Z elements.

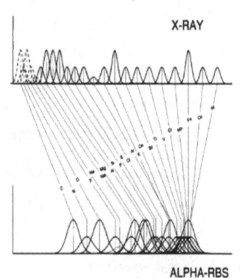

Fig. 5b. The X-ray mode of the APXS separates better the heavier Z elements while the alpha mode has much better resolution power for the light elements. The combination of these two modes makes the APXS a very powerful analytical instrument.

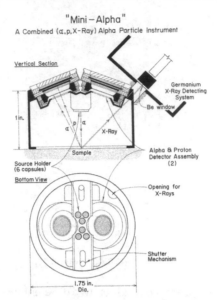

Fig. 6. An Alpha Proton X-ray Spectrometer "Mini-alpha" combining the alpha, proton and x-ray modes in one low power, low volume but high performance analytical instrument on basis of which several version have been used in many planetary missions. The big challenge was to replace the LN cooled x-ray detector with one able to operate at the ambient temperatures.

Based on the "Mini-Alpha" instrument several APXS instruments have been proposed and selected for the following space missions:
1.   Soviet Phobos 1 and Phobos 2 missions to martian satellite Phobos in 1988.
2.   Russian Mars-96 mission to Mars in 1994 (1996)
3.   Pathfinder NASA mission to Mars in 1996
4.   ESA's Rosetta mission to Comet Cheryumov-Gerasimenko in 2002
5.   APXS on NASA Mars Exploration Rover missions (Spirit and Opportunity) in 2003
6.   APXS on NASA Mars Science Laboratory in 2011.

3. Mars pathfinder alpha proton X-ray spectrometer (APXS)

The Alpha Proton X-ray Spectrometer (APXS) for the Mars Pathfinder mission in 1996 [12] was based on the design of the "Mini-Alpha" APXS [8] and used about 45 millicuries of ^{244}Cm alpha source instead the previously used ^{242}Cm isotope. Fig.7 is a photograph of the Pathfinder APXS flight instrument. It consists of two parts: 1/ the sensor head containing nine ^{244}Cm alpha sources in a ring-type geometry and three detectors for the measurement of the three components: A telescope of two Si-detectors for the measurement of alpha-particles and protons and a Si-PIN X-ray detector with its preamplifier, and 2/the main electronic box that contains all the necessary electronic components of a spectrometer and its interface with the spacecraft. The total weight of the Pathfinder APXS is 570 grams and it needed only 370 mWatts of power for its operation. The sensor head of the APXS is mounted in rear of the Sojourner microrover and it is deployed to the surface by its deployment mechanism as it is shown in Figure 8.

The Mars Pathfinder APXS performed very well on Mars and provided us for the first time the chemical composition of martian rocks [13-15]. It was also found that the soil chemical composition was very similar to that found by Viking XRF spectrometer in 1976[15].

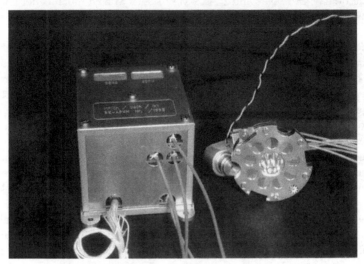

Fig. 7. A photograph of the Pathfinder APXS flight instrument

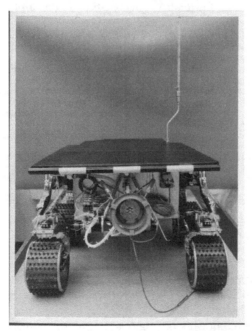

Fig. 8. APXS is mounted on the rear of Sojourner microrover and it is deployed to the surface by its own deployment mechanism.

3.1 Cm-244 alpha sources

For longer mission to distant planetary bodies, [242]Cm has too short a half-life decay time (163 days) and it would be too weak by the time the spacecraft reaches the target planets. [244]Cm with a half-life time of 18.1 years and E_α=5.8 MeV seems to be a much better choice for such missions because it has similar characteristics of 242Cm but has a much longer half-life time. The disadvantage of a longer half-life isotope is that for the same intensity one must pack more radioactive material per unit area, which deteriorates the energy resolution of the source by self absorption in its thickness of the material. Earlier, the [244]Cm was readily available in the US from the Livermore National Laboratory, but presently it is available only from the State Scientific Centre Research Institute of Atomic Reactors Dimitrovgrad, Russia.[21] There, the sources are prepared by high temperature condensation of metal curium vapor onto silicon substrates. For source production, the initial fraction of [244]Cm content was about 93%. Then, before the stage of source preparation, the curium was purified from daughter [240]Pu nuclide and additionally purified from americium, microquantities of [252]Cf and other impurities. The final deposition was on the silicon disks with [244]Cm fixed on their surfaces as a silicide. The sources for the Mars Pathfinder mission have overall dimensions as follows: disk diameter 8 mm; thickness 0.3 mm; and 6 mm diameter active spot. The source activities are 5 ± 1 mCi and the alpha-line half-widths are equal to (1.7–2.5) and (2.9–4.5)% of full width at half maximum energy of 5.8 MeV. The sources and their spectral characteristics stability were studied at wide intervals of physical and chemical parameters of the environment simulating real conditions of storage and maintenance. Thermo–vacuum (from -60 °C up to 1000 °C), mechanical, and vibrational tests were performed to demonstrate that the sources maintained their characteristics. Table 2 lists the main characteristics of the alpha sources used for the Pathfinder APXS instrument.

Radioisotope	[244]Cm
Number of Sources	9
Total Intensity	50 millicuries (1.85*109 Becq.)
Ea	5.807 MeV
$T_{1/2}$	18.1 years
FWHM	2.3 %
FW0.1M	3.5%
FW0.01M	10.0%

Table 2. Characteristics of the [244]Cm alpha radioactive source used for the Mars Pathfinder APXS instrument.

Also, an important factor in determining the resulting energy spread of a source is the chemical composition of the source material: the ideal case is to use the source material in elemental form. In the case of curium, however, the metal becomes chemically unstable and the sources deteriorate rapidly. More recent work [21] concentrated on the formation of curium silicides on the surface of semiconductor grade silicon. This technology has yielded the best results so far and sources for the current or future APXS instruments will be produced by this technique.

The ^{242}Cm and ^{244}Cm alpha sources as used for the lunar and earlier Mars missions are called "open" alpha sources without any cover in order not to degrade the energy resolution of the alpha sources, which is one of the requirements for a good alpha spectrometer. However, using "open" sources present some challenges in handling radioactive sources and preventing source contamination of the spectrometer and its environment. During the fission decay, some recoil product can get out of the source housing and contaminate the instrument and the targets as well. In these cases, a very thin film of Al_2O_3 and VYNS (polypropylene) combination of a total thickness of 1200 Å was used in front of the source collimators. This thickness did not not affect the energy resolution of the alpha particles much but it is thick enough to stop the recoil products and prevent any contamination.

For the most recent APXS, the 244Cm are placed in a sealed housing covered with a thin 2 micron Ti foil. This somewhat degrades the alpha source energy resolution, however without degrading significantly the analytical performance of the APXS. First, for convenience, aluminum foils have been used for that purpose, but their use was discontinued when it was realized that the aluminum is easily oxidizing in the air with time causing continuous decreasing of the energy of the alpha particles.

The big advantage of sealed alpha sources is in much easier handling of the radioactive sources, easier way to prevent source contamination, but mainly it makes the transportation of radioactive material much easier and without a requirement for a special transport containers approved by the IAEA.

4. The Mars exploration rover APXS

For the Mars Rover Exploration (MER) mission in 2003 a more advanced x-ray sensor was developed which uses a silicon drift x-ray detector with a 5-μm beryllium entrance window and has an energy resolution of about 155 eV at 5.49 keV that rivals the energy resolution of the best terrestrial XRF laboratory spectrometers [16]. The rest of the APXS is almost similar to the one used on the Pathfinder. The sensor head of the APXS that contains the alpha sources and all the detectors together with the first stage of amplification electronics was mounted on the Instrument Deployment Device (IDD) of the MER rovers Spirit and Opportunity (robotic arm) and could be deployed to any selected target on the martian surface in sequence with other analytical instruments on the robotic arm, as can be seen in Fig. 9. The rest of the APXS electronics is located inside the temperature controlled compartment of the rover.

The APXS on the MER mission performed very well throughout the entire mission. Although the MER mission was designed to operate only for 3 months on the surface of Mars, we are now still operating successfully after 7 years on the surface of Mars without any degradation in its performance. Actually, due to the decay of radioactive sources used on Mössbauer instrument that was raising the APXS background, the signal to noise ratio in the APXS is better now, 7 years later in the mission, rather than at the beginning of the MER mission in 2004.

Fig. 10 shows the spectra obtained by the APXS on Mars Explorer mission of soil on two different landing sites, on Meridiani planum and Gusev crater. As it can be seen from the spectra, except for some small differences, the soil composition on both sites is very similar. The APXS has analyzed hundreds of martian samples and found many different lithologies, starting with basalt-like rocks, altered basalts, to rocks containing large amount sulfur, phosphor, magnesium, etc., on both landing sites. One such lithology with almost half of the sample in the form of sulfates is shown in Figure 11.

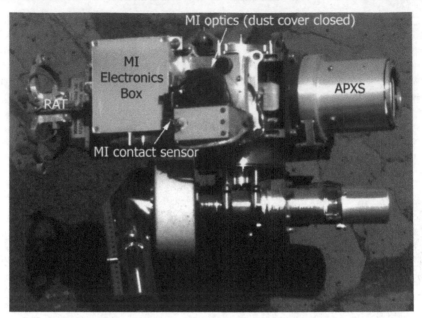

Fig. 9. The Instrument Deployment Device (IDD) of the MER rovers Spirit and Opportunity (robotic arm) showing the mounting of the APXS. The deployment of the APXS to the surface is done by commanding whenever it is desirable.

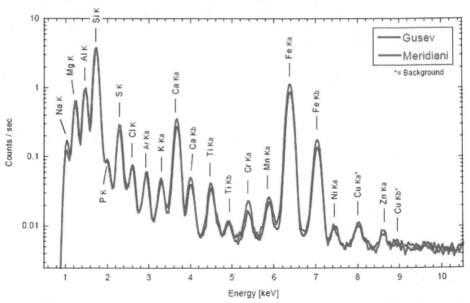

Fig. 10. APXS X-ray spectra of Martian soils measured at Gusev Crater (blue) and Meridiani Planum (red).

Sample Name:	A401_sulf_soil
Spacecraft:	MER2
Sol #:	401
Measurement Start (LST):	05:58:54
Measurement Duration:	6:50:10
DAT.-file #:	2A161961338EDRA600N1438N0M1
LBL.-file #:	2A161961338EDRA600N1438N0M1
Time created:	2005-02-19 06:59:26

Fe^{2+}/Fe_{tot} from MB:	0.7
Relative Geometric Yield :	0.85
Ratio of Pu-Lα scatter lines (coh/incoh):	0.62
Intensity of Pu-Lα coh. scatter line (cps):	0.40
Geometric Yield from scatter lines:	0.67

Ratios:	Mg/Si	Al/Si	FeO/MgO[4]
by wt	0.43	0.29	3.00

APXS Preliminary Analysis

Elements	wt%	Oxides	wt%	mol%
Na[3]	1.60	Na_2O	2.20	2.60
Mg	4.00	MgO	6.60	11.90
Al	2.70	Al_2O_3	5.10	3.60
Si	9.20	SiO_2	19.70	23.90
P	2.53	P_2O_5	5.80	2.97
S	12.30	SO_3	30.70	27.90
Cl	0.45	Cl	0.45	0.90
K	0.07	K_2O	0.09	0.07
Ca	4.60	CaO	6.40	8.30
Ti	0.27	TiO_2	0.44	0.40
Cr	0.04	Cr_2O_3	0.05	0.02
Mn	0.17	MnO	0.21	0.22
Fe	15.50	$(FeO)^{(2)}$	14.00	14.10
Ni (ppm)	70	$(Fe_2O_3)^{(2)}$	6.60	3.00
Zn (ppm)	90	Fe_2O_3T	22.20	
Br (ppm)	460	Σ Oxides[1]	99.9	100.0

Notes:

(1) sample matrix (for matrix correction and geometric yield) by closure, assuming all Fe as Fe_2O_3, and matrix free of H_2O, CO_2, etc.

(2) calculated with Fe^{2+}/Fe_{tot} from MB (default assumption: Fe^{2+}/Fe_{tot} =0.7).

(3) Na too low (by ~30 %) due to faulty algorithm. Will be corrected asap.

(4) assumes all Fe as FeO

Fig. 11. The chemical analyses results from a soil sample A_0401, obtained by the APXS on Spirit on Martian sol 401 showing very high concentration of sulfates.

5. The nano alpha-X instrument for MUSES-C mission

In the year of 2000, NASA, with collaboration with Japanese JAXA, were getting ready for the MUSES-C mission to bring back to Earth a sample from an asteroid. On that mission, it was also proposed to study in-situ the surface characteristics of that asteroid with a nanorover populated with several analytical instruments, including an Alpha-X Spectrometer (AXS) [22] to obtain the chemical composition of the asteroid surface material. For that mission, a further miniaturization of the AXS was required to fit it inside the Jet Propulsion Laboratory's nanorover with the overall dimensions of 14x15x6 cm. A massive hybridization and elimination of the proton mode resulted in a miniaturized Alpha-X Spectrometer shown in Fig. 12 and with specifications listed in Table 3. The AXS, although drastically miniaturized, is a complete spectrometer capable of providing highly accurate analytical results, similar to other APXS instruments.

Fig. 12. A photograph of AXS laboratory instrument for MUSES-C mission.

APXS Characteristics for MUSES-C Mission

Weight:	95 g
Volume: a/ Sensor head	15 cm^3
b/ Electronics	<u>50</u> cm^3
Total Volume	65 cm^3
Power :	<200 mW (~25 mA @7.5 V)
Voltages:	+7.5 V DC
	-7.5 V DC
Radioactive Sources:	30 mCi of Cm-244 radioisotope
E_α:	5.8 MeV
$T_{\frac{1}{2}}$:	18.1 years
Accumulation Time:	3-5 hours for each sample
Data Requirements:	100 kb per sample

Table 3. Characteristics of the AXS instrument for MUSES-C mission.

The MUSES-C mission suffered several long delays and NASA eventually cancelled its participation on this mission. The Japanese, however, went ahead with the mission that was renamed Hayabusa after the launch in 2003 and succeeded in bringing back a small amount of asteroid material from asteroid Itokawa in June, 2010. Despite the cancellation, the AXS was designed, built and fully flight qualified. It is now available for some potential future mission.

6. Summary

We have described here some of the techniques and analytical instruments that have been developed in the past half century and used for space applications. Many other space analytical instruments have used radioactive sources in one way or another that have not been described here. Some of the developed techniques had a profound influence in developing analytical instruments that are used today in many terrestrial laboratories around the world.

Table 4 shows some of the radioisotopes used in different analytical instruments on space missions. Besides the ^{242}Cm and ^{244}Cm that were used in the APXS, the Mössbauer experiment on the MER mission used about 400 millicuries of ^{57}Co at the start of the mission, to obtain the mineralogy of the iron bearing rocks [10]. Similarly, ^{55}Fe has been used in an XRF instrument on Viking mission in 1976 [11] and on the Beagle 2 mission in 2003 [12].

	Isotope	Half Live Time	Energy, Type	Space Mission
1.	Cm-242	161 days	6.1 MeV, α	ASI Surveyor Lunar Missions, 1967-1968
2.	Cm-244	18.1 years	5.8 MeV, α	Phobos1@2('88), Mars96('96), Pathfinder('97), MER(2004), MSL(2011), Rosetta (2002)
3.	Fe-55	2.7 y	5.9 keV, x	Viking 1&2 (1976), Beagle 2 (2003)
4.	Co-57	271 days	14, 122 keV, γ	Viking 1&2 (1976), MER (2004)
5.	Am-241	432 y	5.49 MeV,α,γ	XRF
6.	Pu-238	87.74 y	5.5 MeV, α	Radioisotope Thermoelectric Generators (RTG), Radioactive Heating Units (RHU), many space missions
7.	Pu-239	2.4x104 y	5.15 MeV,α	Calibrations

Table 4. The most common radioisotopes used in the analytical instrument for space applications

7. References

[1] Turkevich, A., "Chemical analysis of surfaces by use of large-angle scattering of heavy charged particles", Science, 134, 672, 1961.

[2] A. L. Turkevich, W. A. Anderson, T. E. Economou, E. J. Franzgrote, H. E. Griffin, S. L. Crotch, J. H. Patterson and K. P. Sowinski. "The Alpha-Scattering Chemical Analysis Experiment on the Surveyor Lunar Missions". Jet Propulsion Laboratory Technical Report 52-1265, pp. 505-82, June 15, 1968.

[3] Patterson, J.H, A. L. Turkevich and E. J. Franzgrote, Chemical analysis of surfaces using alpha particles, J. Geophys.Res., 70, 1311, 1965.

[4] Thanasis E. Economou, Anthony L. Turkevich, Keneth P. Sowinski, James H. Patterson and Ernest J. Franzgrote, The Alpha-Scattering Technique of Chemical Analysis, J. Geophys. Res., 75, No 32, 6514, 1970.

[5] Thanasis E. Economou,Anthony L. Turkevich and James H. Patterson, An Alpha Particle Experiment for Chemical Analysis of the Martian Surface and Atmosphere, J. Geophys. Res., 78, No 5, 781, 1973.

[6] Patterson, J.H, Ernest J. Franzgrote,A. L. Turkevich, W.A. Anderson, T.E. Economou, H.E. Griffin, S.L. Groach and K.P. Sowinski, Alpha Scattering Experiment on Surveyor 7: Comparison with Surveyor 5 and 6, J. Geophys. Res., 74, 6120, 1969.

[7] James H. Patterson, Harry E. Griffin, E. Philip Horwitz, and Carol A. Bloomquist, Preparation of High Level Alpha-Particle Sources for the Surveyor Alpha Scattering Experiment, Nuclear Technology, Vol. 18, 277-285, 1073.

[8] Thanasis E. Economou and Anthony L. Turkevich. "An Alpha Particle Instrument with Alpha, Proton and X-ray Modes for Planetary Chemical Analyses". Nucl. Instr. & Methods 134, 1976, p. 391-399.

[9] Hovestadt, D., et al., PHOBOS, Proceedings of the International Workshop, Moscow, 1986, p 302

[10] Economou, T.E, J.S. Iwanczyk and R. Rieder, A HgI_2 X-ray Instrument for the Soviet Mars '94 Mission, Nucl. Instr. &   Methods A322,633-638, 1992.

[11] Linkin, V., et al., "A Sophisticated Lander for Scientific   Exploration of Mars : Scientific Objective and Implementation", Planetary Space Science Journal, Vol. 46, Issues 6-7, 1998, pp 717-737.

[12] Rieder, R., H. Wänke, T. Economou and A. Turkevich; "Determination of the Chemical Composition of Martian Soil and Rocks: The Alpha-Proton-X-Ray Spectrometer", J. Geophys. Res. 1997 pp. 4027-4044.

[13] Rieder, R., T. Economou, J. Brückner, G. Dreibus, H. Wänke and A. Turkevich, First Elemental Analysis of Martian Surface by the Mobile Alpha Proton X-ray Spectrometer Attached to Mars Pathfinder Rover Sojourner. Meteoretics and Planetary Science, Vol. 32, No. 4, A107, 1997.

[14] Rieder, R., T. Economou, H. Wänke, A. Turkevich, J. Crisp, J. Brückner, G. Dreibus,H.Y. McSween, Jr., The Chemical Composition of Martian Soil and Rocks Returned by the Mobile Alpha Proton X-ray Spectrometer: Preliminary Results from the X-ray Mode, Science Vol. 278, (1997)1771-1774.

[15] C. Nicole Foley, Thanasis E. Economou and Robert N. Clayton, Final Chemical Results from the Mars Pathfinder Alpha Proton X-ray Spectrometer, J. Geoph. Res. Vol. 108, No 12, doi:10.1029/2003JE002019, 2003.

[16] Rieder, R., et al., *J. Geophys. Res., 108(E12), doi:10.1029/2003JE 002150, 2003*

[17] http://www.esa.int/esaMI/Rosetta

[18] Klingelhoefer, G., J. Geophys. Res., 108, p. 8067, 2003

[19] Clark,B.C., et al., J. Geophys. Res., 87, p.10059, 1982

[20] Beagle 2 mission to Mars in 2003, http://www.beagle2.com/index.htm

[21] V. Radchenko, B. Andreichikov, H. Wanke, V. Gavrilov, B. Korchuganov, R. Rieder, M Ryabinin, and T. Economou, Curium-244 alpha-sources for space applications, Applied Radiation and Isotopes, 2000 Oct., 53(4-5):821-824.

[22] Economou, T.E, The Chemical Composition of an Asteroid Surface by the Alpha X-Ray Spectrometer on the MUSES-C Mission Lander, In Lunar and Planetary Science XXXI, 2000, 1861.pdf

9

The Newly Calculations of Production Cross Sections for Some Positron Emitting and Single Photon Emitting Radioisotopes in Proton Cyclotrons

E. Tel[1], M. Sahan[1], A. Aydin[2], H. Sahan[1], F. A.Ugur[1] and A. Kaplan[3]
[1] Faculty of Arts and Science, Osmaniye Korkut Ata University,
[2] Faculty of Arts and Science, Kirikkale University
[3] Faculty of Arts and Science, Süleyman Demirel University
Turkey

1. Introduction

Nowadays, radioisotopes are produced using both nuclear reactors and cyclotrons. Especially, the induced by intermediate and high energy protons nuclear reactions are very important because of a wide range technical applications. These reactions are required for advanced nuclear systems, such as spallation reaction for production of neutrons in spallation neutron source (capable of incinerating nuclear waste and producing energy), high energy proton induced fission for the radioisotope production alternatives etc. [1,2]. By using the intermediate proton induced reactions, we can directly produce radionuclides used in medicine and industry.

In the last decade, a big success has been provided on production and usage of the radionuclides. The radioisotopes obtained from using charged particles (proton, deuteron, alpha etc.) play an important role in medical applications [3-6]. A medical radioisotope can be classified as a diagnostic or a therapeutic radionuclide, depending on its decay properties. Radionuclides are used in diagnostic studies via emission tomography, i.e. Positron Emission Tomography (PET), Single Photon Emission Computed Tomography (SPECT), and Endoradiotheraphy (internal therapy with radio nuclides). In general, the diagnostic radioisotopes can also be classified into two groups; namely β^+-emitters (^{11}C, ^{13}N, ^{15}O, ^{18}F, ^{62}Cu, ^{68}Ga, etc.) and γ - emitters (^{67}Ga, ^{75}Se, ^{123}I, etc.). The use of positron emitting radioisotopes such as ^{11}C, ^{13}N, ^{15}O, and ^{18}F together with PET offers a highly selective and quantitative means for investigating regional tissue biochemistry, physiology and pharmacology [7]. The positron emitting nuclei which are neutron deficient isotopes are important for PET studies. Positrons annihilate with electrons emitting two photons (Eγ=511 keV) in opposite direction. Most of the positron emitters are still being studied in terms of their applicability for diagnostic purposes. PET has been developing with the increasing number of clinical facilities raising interest in the use of PET in routine practice [8,9].

In the radioisotope production procedure, the nuclear reaction data are mainly needed for optimization of production rates. This process involves a selection of the projectile energy range that will maximize the yield of the product cross section for nuclear reaction and minimize that of the radioactive impurities [5,6]. The total cross section of production yields are also important in accelerator technology from the point of view of radiation protection safety. The nuclear reaction calculations based on standard nuclear reaction models can be helpful for determining the accuracy of various parameters of nuclear models and experimental measurements. Today, experimental cross-sections are available in EXFOR file [7]. The theoretical calculations of production rates for different medical nucleus reactions were calculated in statistical equilibrium (compound) and pre-equilibrium model in literature [10-14].

In this study, the newly calculations of proton cyclotron production cross sections in some PET , SPECT and others used in medical applications radioisotopes used in medical applications were investigated in a range of 5-100 MeV incident proton energy range. Excitation functions for pre-equilibrium calculations were newly calculated by using hybrid model, geometry dependent hybrid (GDH) model. The reaction equilibrium component was calculated with a traditional compound nucleus model developed by Weisskopf-Ewing (W-E) model [16]. We have investigated the optimum energy range for proton cyclotron production cross sections. We have presented the decay data taken from the NUDAT database [23]. Calculated results were also compared with the available excitation functions measurements in EXFOR file [7]. The optimum energy range and the decay data for the investigated radionuclides are given in Table 1.

2. The nuclear reaction cross-section calculations

Calculations based on nuclear reaction models play an important role in the development of reaction cross sections [15]. For many years, it has been customary to divide nuclear reactions into two extreme categories. Firstly, there are very fast, direct reactions which on a time scale comparable to the time $(\cong 10^{-22}$ s) necessary for the projectile to traverse a nuclear diameter, involve simple nuclear excitations, and are non-statistical in nature. Secondly, there are equilibrium nucleus reactions which occur on a very much longer time scale $(\cong 10^{-16}$ to 10^{-18}s) where emissions can be treated by the nuclear statistical model. This second process can be described adequately with equilibrium nucleus theories developed by the Weisskopf-Ewing (W-E) [16] and Hauser-Feshbach [17]. Equilibrium nucleus wave function is very complicated, involving a large number of particle-hole excitations to which statistical considerations are applicable. The spectra of the emitted particles of equilibrium nucleus are approximately Maxwellian, and angular distributions of emitted particles are symmetric about 90 degrees. During the nineteen-fifties and sixties, evidence accumulated suggesting that in some nuclear reactions, it is not possible to understand all emission processes in terms of equilibrium nucleus and direct processes. This reaction is known as pre-equilibrium reaction. The pre-equilibrium reactions occur on time scale about 10^{-18} to 10^{-20} s. Deviations from a Maxwellian shape for the emission spectra were observed for intermediate to high emission energies, with the theory under predicting data. The first developments were made to understand these observations by Griffin [18], who proposed the pre-equilibrium 'exciton model'. Pre-equilibrium processes are important mechanisms in nuclear reactions induced by light projectiles with incident energies above about 8-10 MeV. After Griffin introduced the exciton model, a series of semi-classical models [19-21] of varying complexities have been developed for calculating and evaluating particle emissions in the continuum. More recently, researchers have formulated several quantum-mechanical reaction theories [22] that are based on multi-step concepts and in which

The Newly Calculations of Production Cross Sections for Some Positron Emitting and Single Photon
Emitting Radioisotopes in Proton Cyclotrons

161

statistical evaporation at lower energies is connected to direct reactions at higher energies. The
hybrid model for pre-compound decay is formulated by Blann [19] as

$$\frac{d\sigma_\upsilon(\varepsilon)}{d\varepsilon} = \sigma_R P_\upsilon(\varepsilon)$$

$$P_\upsilon(\varepsilon)d\varepsilon = \sum_{\substack{n=n_0 \\ \Delta n=+2}}^{\bar{n}} \left[{}_n\chi_\upsilon N_n(\varepsilon,U) / N_n(E)\right] g\, d\varepsilon[\lambda_c(\varepsilon) / (\lambda_c(\varepsilon) + \lambda_+(\varepsilon))]D_n \qquad (1)$$

where $P_\upsilon(\varepsilon)d\varepsilon$ represents number of particles of the υ type (neutron or proton) emitted into
the unbound continuum with channel energy between ε and $\varepsilon + d\varepsilon$. The quantity in the first
set of square brackets of Eq.(1) represents the number of particles to be found (per MeV) at a
given energy (with respect to the continuum) for all scattering processes leading to an "n"
exciton configuration. It has been demonstrated that the nucleon-nucleon scattering energy
partition function Nn(E) is identical to the exciton state density ρn(E). The second set of
square brackets in Eq. (1) represents the fraction of the υ type particles at energy, which
should undergo emission into the continuum, rather than making an intra-nuclear
transition. The Dn represents the average fraction of the initial population surviving to the
exciton number being treated. Early, $\lambda_c(\varepsilon)$ is emission rate of a particle into the continuum
with channel energy ε and $\lambda_+(\varepsilon)$ is the intranuclear transition rate of a particle. Nn (ε ,U)
is the number of ways. comparisons between experimental results, pre-compound exciton
model calculations, and intra-nuclear cascade calculations indicated that the exciton model
gave too few pre-compound particles and that these were too soft in spectral distribution for
the expected initial exciton configurations. The intra-nuclear cascade calculation results
indicated that the exciton model deficiency resulted from a failure to properly reproduce
enhanced emission from the nuclear surface. In order to provide a first order correction for
this deficiency the hybrid model was reformulated by Blann and Vonach [24]. In this way
the diffuse surface properties sampled by the higher impact parameters were crudely
incorporated into the pre-compound decay formalism, in the geometry dependent hybrid
(GDH) model. The differential emission spectrum is given in the GDH model as

$$\frac{d\sigma_\upsilon(\varepsilon)}{d\varepsilon} = \pi \lambda^2 \sum_{l=0}^{\infty}(2l+1)T_l\, P_\upsilon(l,\varepsilon) \qquad (2)$$

where λ is reduced de Broglie wavelength of the projectile and T_l represents transmission
coefficient for l th partial wave ℓ is orbital angular momentum in n unit $\hbar$.

3. Results and discussion

This work describes new calculations on the excitation functions of $^{18}O(p,n)^{18}F$, $^{57}Fe(p,n)^{57}Co$,
$^{57}Fe(p,\alpha)^{54}Mn$, $^{68}Zn(p,2n)^{67}Ga$, $^{68}Zn(p,n)^{68}Ga$, $^{93}Nb(p,4n)^{90}Mo$, $^{112}Cd(p,2n)^{111}In$, $^{127}I(p,3n)^{125}Xe$
,$^{133}I(p,6n)^{128}Ba$ and $^{203}Tl(p,3n)^{201}Pb$ reactions carried out in the 5-100 MeV proton energy range.
The pre-equilibrium calculations involve the hybrid model and the geometry dependent
hybrid (GDH) model. Equilibrium reactions have been calculated according to Weisskopf-
Ewing (W-E) model. The ALICE/ASH code was used in the calculations of all models
described above. The ALICE/ASH code is an advanced and modified version of the ALICE

codes. The modifications concern the implementation in the code of models describing the pre-compound composite particle emission, fast γ-emission, different approaches for the nuclear level density calculation, and the model for the fission fragment yield calculation. The ALICE/ASH code can be applied for the calculation of excitation functions, energy and angular distribution of secondary particles in nuclear reactions induced by nucleons and nuclei with the energy up to 300 MeV. The initial exciton number as n_o =3 and the exciton numbers (for protons and neutrons) in the calculations for proton induced reactions as,

$$_3X_p = 2\frac{\left(\sigma_{pn}/\sigma_{pp}\right)N + 2Z}{2(\sigma_{pn}/\sigma_{pp})N + 2Z}, \quad _3X_n = 2 - {_3X_p} \tag{3}$$

where σ_{xy} is the nucleon-nucleon interaction cross-section in the nucleus. Z and N are the proton and neutron numbers, respectively, of the target nuclei. The ratio of nucleon-nucleon cross-sections calculated taking into account to Pauli principle and the nucleon motion is parameterized

$$\sigma_{pn}/\sigma_{pp} = \sigma_{np}/\sigma_{nn} = 1.375 \times 10^{-5}T^2 - 8.734 \times 10^{-3}T + 2.776 \tag{4}$$

where T is the kinetic energy of the projectile outside the nucleus. The super-fluid model [26] has been applied for nuclear level density calculations in the ALICE/ASH code. The results of the calculations are plotted in [Fig. 1], [Fig. 2], [Fig. 3], [Fig. 4], [Fig. 5], [Fig. 6], [Fig. 7], [Fig. 8], [Fig. 9], and [Fig. 10]. Calculated results based on hybrid model, geometry dependent hybrid model and equilibrium model have been compared with the experimental data. The experimental taken from EXFOR [7] are shown with different symbols (such as +, o, Δ) symbols in all figures. The results are given below.

3.1 ^{18}O(p,n)^{18}F reaction process
The calculation for the excitation function of ^{18}O(p,n)^{18}F reaction has been compared with the experimental values in Fig.1. In general, the hybrid model calculations are the best agreement with the measurements of ^{18}O(p,n)^{18}F reaction up to 30 MeV the incident proton energy. The equilibrium W-E model calculations are only in agreement with in energy region lower than 20 MeV. The optimum energy range for production of ^{18}F is E_p= 10 → 5 MeV.

3.2 ^{57}Fe(p,n) ^{57}Co reaction process
The calculation cross section of ^{57}Fe(p,n) ^{57}Co reaction has been compared with the experimental values in Fig.2. The hybrid model calculations are the best agreement with the measurements of Levkovskij [7] at above the 20 MeV energie regions. The W-E model calculations are only in agreement with in energy region lower than 20 MeV. The GDH model calculations are a little higher than the measurements of Levkovskij [7] at above the 20 MeV for incident proton energies. The optimum energy range for production of ^{57}Co is E_p= 15 → 5 MeV.

3.3 ^{57}Fe(p,α)^{54}Mn reaction process
The calculated ^{57}Fe(p,α)^{54}Mn reaction has been compared with the experimental values in Fig.3. The hybrid model and GDH model calculations are in good agreement with the measurements of Levkovskij [7] at above the 20 MeV energie regions. The W-E model calculations are good in agreement with in energy region lower than 20 MeV. The optimum energy range for production of ^{54}Mn is E_p= 20 → 10 MeV.

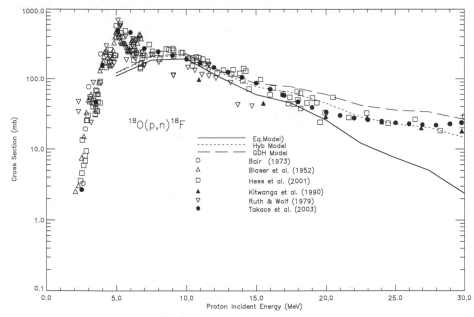

Fig. 1. The comparison of the calculated cross section of $^{18}O(p,n)^{18}F$ reaction with the values reported in Ref. [7].

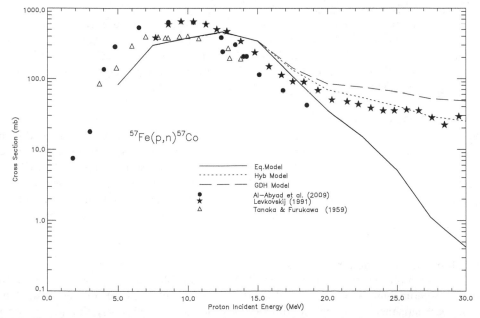

Fig. 2. The comparison of the calculated cross section of $^{57}Fe(p,n)^{57}Co$ reaction with the values reported in Ref. [7].

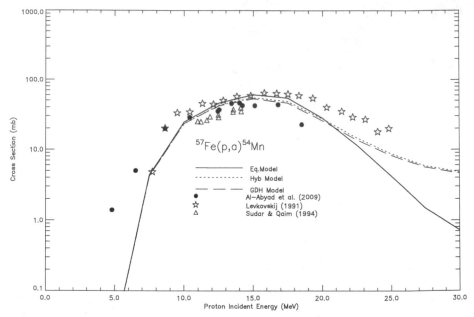

Fig. 3. The comparison of the calculated cross section of $^{57}Fe(p,\alpha)^{54}Mn$ reaction with the values reported in Ref. [7].

3.4 ^{68}Zn(p,2n)^{67}Ga reaction process

The calculation on the excitation function of $^{68}Zn(p,n)^{67}Ga$ reaction has been compared with the experimental values in Fig.4. The W-E model calculations are in agreement with the measurements up to 25 MeV. Also, the GDH and hybrid model calculations are in very good harmony with the experimental data. The optimum energy range for production of ^{67}Ga is $E_p= 30 \rightarrow 15$ MeV.

3.5 ^{68}Zn(p,n)^{68}Ga reaction process

The calculation on the excitation function of $^{68}Zn(p,n)^{68}Ga$ reaction has been compared with the experimental values in Fig.5. The experimental data of the measurements are in good agreement with each other. The W-E model calculations are in agreement with the measurements up to 15 MeV. The GDH model calculations are the best agreement with the experimental data of 5 – 30 MeV energy range. Also, the hybrid model calculations are in very good harmony with the experimental data. The optimum energy range for production of ^{68}Ga is $E_p= 15 \rightarrow 5$ MeV.

3.6 ^{93}Nb(p,4n)^{90} Mo reaction process

The calculation on the excitation function of $^{93}Nb(p,4n)^{90}Mo$ reaction has been compared with the experimental values of Ditroi et al. [7] in Fig.6. While the GDH model calculations are the best agreement with the experimental data of 35 – 60 MeV energy range, the other model calculations are higher than the experimental data. The optimum energy range for production of ^{90}Mo is $E_p= 55 \rightarrow 45$ MeV.

The Newly Calculations of Production Cross Sections for Some Positron Emitting and Single Photon
Emitting Radioisotopes in Proton Cyclotrons

165

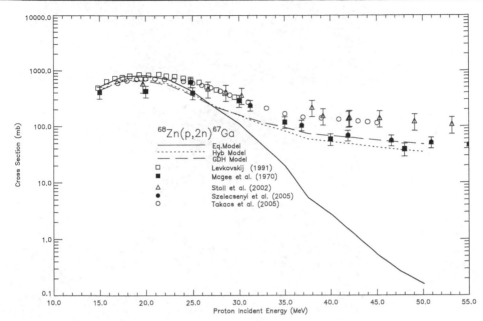

Fig. 4. The comparison of the calculated cross section of ^{68}Zn(p,2n)^{67}Ga reaction with the values reported in Ref. [7].

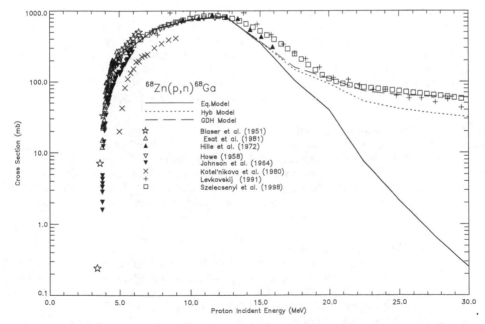

Fig. 5. The comparison of the calculated cross section of ^{68}Zn(p,n)^{68}Ga reaction with the values reported in Ref. [7].

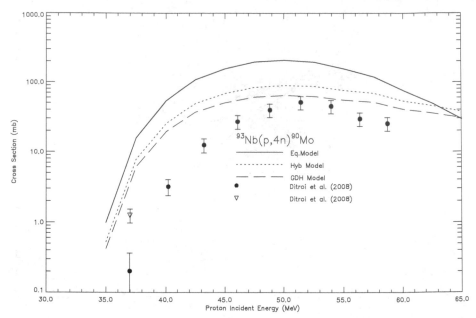

Fig. 6. The comparison of the calculated cross section of ^{93}Nb(p,4n)^{90}Mo reaction with the values reported in Ref. [7].

3.7 ^{112}Cd(p,2n)^{111}In reaction process

Especially, the ^{111}In radionuclei are very important for SPECT. The calculation on the excitation function of ^{112}Cd(p,2n)^{111}In reaction has been compared with the experimental values in Fig.7. Generally, the experimental data of the measurements are in good agreement with each other. The GDH and hybrid model calculations are in very good harmony with the experimental data. The equilibrium W-E model calculations are in agreement with the measurements up to 30 MeV. The optimum energy range for production of ^{111}In is E_p= 25 →15 MeV.

3.8 ^{127}I(p,3n)^{125}Xe reaction process

The calculation on the excitation function of ^{127}I(p,3n)^{125}Xe reaction has been compared with the experimental values in Fig.8. We can say that the experimental data of the measurements are in good agreement with each other. The equilibrium W-E model calculations are in agreement with experimental values up to 35 MeV The GDH and hybrid model calculations are in very good harmony with the experimental data. The optimum energy range for production of ^{125}Xe is E_p= 35 →25 MeV.

3.9 ^{133}Cs(p,6n)^{128}Ba reaction process

The calculation on the excitation function of ^{133}Cs(p,6n)^{128}Ba reaction has been compared with the experimental values in Fig.9. The equilibrium W-E model calculations are not agreement with experimental values. The GDH model calculations are the best agreement with the experimental data of Deptula *et al.* [7] for 45-100 MeV energy range. While the hybrid model calculations are the good agreement with the experimental data of 45-65 MeV energy range, for above the 65 MeV proton incident energy this model calculations are higher than the experimental data.The optimum energy range for production of ^{128}Ba is E_p= 70 →50 MeV.

The Newly Calculations of Production Cross Sections for Some Positron Emitting and Single Photon
Emitting Radioisotopes in Proton Cyclotrons

167

3.10 ^{203}Tl(p,3n)^{201}Pb reaction process

The ^{201}Tl radioisotopes are very important for using in the SPECT. The reaction process for ^{201}Tl is ^{203}Tl(p,3n)^{201}Pb, $^{201}Pb \xrightarrow{9.33h} ^{201}Tl$. The calculation on the excitation function of ^{203}Tl(p,3n)^{201}Pb reaction has been compared with the experimental values of Blue *et al* [7], Lebowitz *et al.* [7] and Al-saleh *et al.* [7] Takacs *et al.* [7] in Fig.8. The Weisskopf-Ewing model calculations are in agreement with the measurements up to 35 MeV. The GDH model and hybrid model calculations are in good agreement with the experimental data of 20 – 35 MeV energy range. The optimum energy range for production of ^{201}Tl is E_p= 30→ 20 MeV.

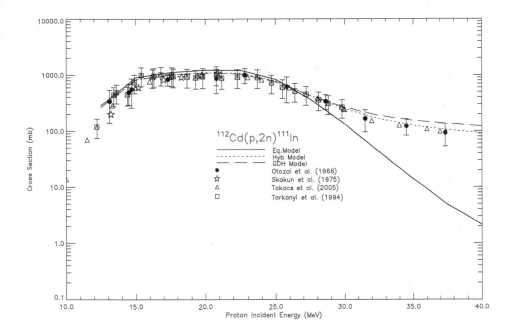

Fig. 7. The comparison of the calculated cross section of ^{112}Cd(p,2n)^{111}In reaction with the values reported in Ref. [7].

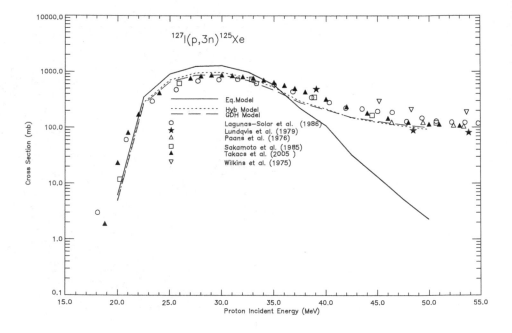

Fig. 8. The comparison of the calculated cross section of ^{127}I(p,3n)^{125}Xe reaction with the values reported in Ref. [7].

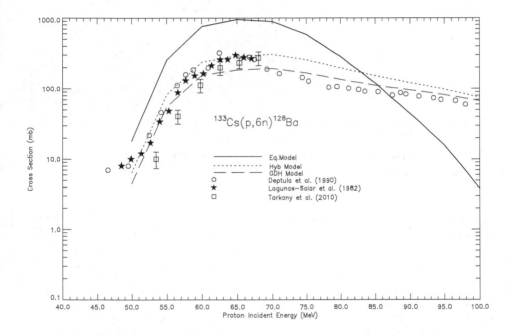

Fig. 9. The comparison of the calculated cross section of ^{133}Cs(p,6n)^{128}Ba reaction with the values reported in Ref. [7].

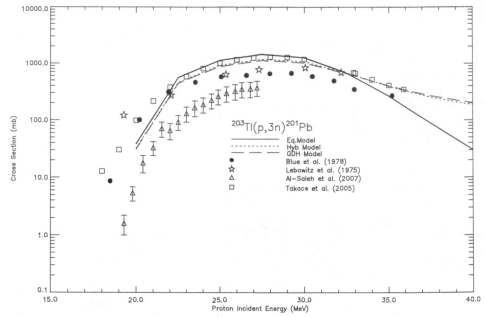

Fig. 10. The comparison of the calculated cross section of ^{203}Tl(p,3n)^{201}Pb reaction with the values reported in Ref. [7].

Producted radioisotopes	Half life	Mode of decay (%)	$E_\Box$ (keV)	$I_{\Box\Box}$(%)	Optimum Energy Range (MeV)
^{18}F	1.83 h	EC + β⁺ (100)	0.52	0.01795	E_p= 10 → 5
^{57}Co	271.74 d	EC + β⁺ (100)	122.06065	85.60	E_p= 15 → 5
^{54}Mn	312.05d	EC + β⁺ (100)	834.848	99.97	E_p= 20 → 10
^{67}Ga	3.2617 d	EC + β⁺ (100)	393.527	4.56	E_p= 30 →15
^{68}Ga	67.63m	EC+ β⁺ (100)	1077.35	3	E_p= 15 → 5
^{90}Mo	5.67h	EC + β⁺ (100)	257.34	78	E_p= 55 →45
^{111}In	2.8047 d	EC + β⁺ (100)	245.350	94.10	E_p= 25 →15
^{125}Xe	56.9s	EC + β⁺ (100)	111.3	60.2	E_p= 35 →25
^{128}Ba	2.43d	EC + β⁺ (100)	273.44	14.5	E_p= 70 →50
^{201}Pb	9.33 h	EC+ β⁺ (100)	546.280	0.279	E_p= 30→ 20

Table 1. The decay data and optimum energy range for investigated radionuclides.

The Newly Calculations of Production Cross Sections for Some Positron Emitting and Single Photon
Emitting Radioisotopes in Proton Cyclotrons

171

4. Conclusions

The new calculations on the excitation functions of $^{18}O(p,n)^{18}F$, $^{57}Fe(p,n)^{57}Co$, $^{57}Fe(p,\alpha)^{54}Mn$, $^{68}Zn(p,2n)^{67}Ga$, $^{68}Zn(p,n)^{68}Ga$, $^{93}Nb(p,4n)^{90}Mo$, $^{112}Cd(p,2n)^{111}In$, $^{127}I(p,3n)^{125}Xe$, $^{133}I(p,6n)^{128}Ba$ and $^{203}Tl(p,3n)^{201}Pb$ reactions have been carried out using nuclear reaction models. Although there are some discrepancies between the calculations and the experimental data, in generally, the new evaluated hybrid and GDH model calculations (with ALICE/ASH) are good agreement with the experimental data above the incident proton energy with 5-100 MeV in Figs. 1-10. While the Weisskopf-Ewing model calculations are only in agreement with the measurements for lower incident proton energy regions, hybrid model calculations are in good harmony with the experimental data for higher incident proton energy regions. Some nuclei used in this study were examined and compared in previous paper written by Tel et al.[13,14]. Detailed informations can be found in these papers. And also new developed semi-empirical formulas for proton incident reaction cross-sections can be found in Ref. [27,28].

When Comparing the experimental data and theoretical calculations, the production of ^{18}F,^{57}Co, ^{54}Mn, $^{67,68}Ga$, ^{90}Mo, ^{111}In, ^{125}Xe , ^{128}Ba and ^{201}Pb radioisotopes can be employed at a medium-sized proton cyclotron since the optimum energy ranges are smaller than 50 MeV, except for ^{128}Ba. We gave the optimum energy range and the decay data for the investigated radionuclides in Table 1.

5. References

[1] M B Chadwick, P G Young, S Chiba, S C Frankle, G M Hale, H G Hughes, A J Koning, R C Little, R E MacFarlane, R E Prael, L S Waters, *Nucl. Sci. Engin.* 131 293 (1999)

[2] C. Rubbia, J A Rubio, S Buorno, F Carminati, N Fitier, J Galvez, C Gels, Y Kadi, R Klapisch, P Mandrillon, J P Revol, and Ch. Roche, *European Organization for Nuclear Research*, CERN/AT/95-44 (ET) (1995).

[3] S M Qaim *Radiat. Phys. Chem.* 71 917 (2004).

[4] S M Qaim *Radiochim. Acta* 89 297 (2001).

[5] B Scholten, E Hess, S Takacs, Z Kovacs, F Tarkanyi, H H Coenen and S M Qaim *J. Nucl. Sci. and Tech.* 2 1278 (2002).

[6] S M Qaim *Radiochim. Acta* 89 223 (2001).

[7] EXFOR/CSISRS (Experimental Nuclear Reaction Data File), Brookhaven National Laboratory, National Nuclear Data Center, (http://www.nndc.bnl.gov/exfor/) (2009) .

[8] A P Wolf, J S Fowler *Positron Emitter Labeled Radiotracers, Chemical Considerations in Positron Emission Tomography* (Alan R. Liss, Inc. Pub.) (1985) .

[9] A P Wolf and W B Jones *Radiochim. Acta* 34 1 (1983).

[10] K Gul *Appl. Radiat. Isotopes* 54 147 (2001).

[11] K Gul *Appl. Radiat. Isotopes* 54 311 (2001).

[12] A Aydın, B Sarer, E Tel *Appl. Radiat. Isotopes* 65 365 (2007).

[13] E Tel, E G Aydin, A. Kaplan and A. Aydin, *Indian J. Phys.* 83 (2) 1-20 (2009).

[14] E G Aydin, E Tel, A Kaplan and A Aydin, *Kerntechnic,* 73, 4, (2008).

[15] M B Chadwick *Radiochim. Acta* 89 325 (2001).

[16] V F Weisskopf and D H Ewing *Phys. Rev.* 57 472 (1940) .

[17] W Hauser and H Feshbach *Phys. Rev.* 87 366 (1952) .

[18] J J Griffin, *Phys. Rev. Lett.* 17 478 (1966) .

[19] M Blann *Annu. Rev. Nucl. Sci.* 25 123 (1975) .
[20] E Betak *Comp. Phys. Com.* 9 92 (1975).
[21] H. Gruppelaar, P Nagel, P E Hodgson and LaRivasta Del *Nuovo Cim.* 9 1 (1986).
[22] H Feshbach, A Kerman and S Koonin *Annu.Phys.* (NY), 125 429 (1980) .
[23] NUDAT – Decay Radiation Database, http://www.nndc.bnl.gov/nudat2
[24] M Blann and H K Vonach *Phys. Rev.* C28 1475 (1983).
[25] C. H. M. Broeders, A. Yu. Konobeyev, Yu. A. Korovin, et al.,
          http://bibliothek.fzk.de/zb/berichte/FZKA7183.pdf.
[26] A. V. Ignatyuk, K. K. Istekov, G. N. Smirenkin, *Yad. Fiz.* 29, 875 (1979).
[27] Tel, E., Aydın, E. G., Aydın, A., Kaplan, A., *Appl. Radiat. Isotopes* 67 (2), 272 (2009)
[28] Tel, E., Aydın, A., Aydın, E. G., Kaplan, A., Ö. Yavaş, İ. Reyhancan, , *Pramana-J. Phys.,*
       74 (6), 931-944, (2010).

Cesium (^{137}Cs and ^{133}Cs), Potassium and Rubidium in Macromycete Fungi and *Sphagnum* Plants

Mykhailo Vinichuk[1,3], Anders Dahlberg[2] and Klas Rosén[1]
[1]Department of Soil and Environment, Swedish University of Agricultural Sciences,
[2]Department of Forest Mycology and Pathology, Swedish University of
Agricultural Sciences,
[3]Department of Ecology, Zhytomyr State Technological University,
[1,2]Sweden
[3]Ukraine

1. Introduction

1.1 Cesium (^{137}Cs and ^{133}Cs), potassium and rubidium in macromycete fungi

Radiocesium (^{137}Cs) released in the environment as result of nuclear weapons tests in the 1950s and 1960s, and later due to the Chernobyl accident in 1986, is still a critical fission product because of its long half-life of 30 years and its high fission yield. The study of the cesium radioisotope ^{137}Cs is important, as production and emission rates are much higher than other radioisotopes. This chapter comprises results obtained in several experiments in Swedish forest ecosystems and aims to discuss the behavior of cesium isotopes (^{137}Cs and ^{133}Cs) and their counterparts potassium (K) and rubidium (Rb) in the "soil-fungi-plants transfer" system. The chapter consists of two parts: one mainly dealing with ^{137}Cs, ^{133}Cs, K and Rb in forest soil and macromycete fungi, and the other with the same isotopes in separate segments of *Sphagnum* plants.

The bioavailability of radionuclides controls the ultimate exposure of living organisms and the ambient environment to these contaminants. Consequently, conceptually and methodologically, the understanding of bioavailability of radionuclides is a key issue in the field of radioecology. Soil-fungi-plants transfer is the first step by which ^{137}Cs enters food chains.

1.1.1 The role of fungi in ^{137}Cs transfer in the forest

The availability of radionuclides (^{137}Cs in particular) in soils of different ecosystems is to a large extent regulated by various vascular plants and fungal species. Thus, the behavior of ^{137}Cs in forest ecosystems differs substantially from other ecosystems, foremost due to the abundance of fungal mycelia in soil, which contribute to the persistence of the Chernobyl radiocesium in the upper horizons of forest soils (Vinichuk & Johanson, 2003). Both saprotrophic and mycorrhizal fungi have key roles in nutrient and carbon cycling processes in forest soils. The mycelium of soil fungi has a central role in breaking down organic matter and in the uptake of nutrients from soil into plants via the formation of symbiotic mycorrhizal associations (Read & Perez-Moreno, 2003). The fungi facilitate nutrient uptake into the host plant, both as a consequence of the physical geometry of the mycelium and by the ability of the fungi to mobilize nutrients from organic substrates through the action of extracellular catabolic enzymes (Leake & Read, 1997). In addition to acquiring essential macronutrients, mycorrhizal fungi are efficient at taking-up and accumulating microelements (Smith & Read, 1997), this ability results in the accumulation of non-essential elements and radionuclides, particularly ^{137}Cs and can have important consequences for the retention, mobility and availability of these elements in forest ecosystems (Steiner et al., 2002).

Although fungal biomass, in comparison to plant biomass, is relatively low in forest soil (Dighton et al., 1991; Tanesaka et al., 1993), many fungal species accumulate more ^{137}Cs than vascular plants do and ^{137}Cs activity concentrations in many fungi are 10 to 100 times higher than in plants (Rosén et al., 2011). Fungi (particularly sporocarps) accumulate ^{137}Cs against a background of low ^{137}Cs activity concentrations, thus, the contribution of fungi to ^{137}Cs cycling in forest systems is substantial.

Fungi are important in radiocesium migration in nutrient poor and organic rich soils of forest systems (Rafferty et al., 1997). In organic matter, the presence of single strains of saprotrophic fungi considerably enhances the retention of Cs in organic systems (Parekh et al., 2008): ≈ 70% of the Cs spike is strongly (irreversibly) bound (remains non-extractable) compared to only ≈ 10% in abiotic (sterilized) systems. Fungal mycelium may act as a sink for radiocesium (Dighton et al., 1991; Olsen et al., 1990), as it contains 20–30% [137]Cs in soil inventories, and as much as 40% of radiocesium can leached from irradiated samples compared to control samples (Guillitte et al., 1994). Mycelium in upper organic soil layers may contain up to 50% of the total ^{137}Cs located within the upper 0-10 cm layers of Swedish and Ukrainian forest soils (Vinichuk & Johanson 2003). In terms of the total radiocesium within a forest ecosystem, fungal sporocarps contain a small part of activity and may only account for about 0.5 % (McGee et al., 2000) or even less – 0.01 to 0.1% (Nikolova et al., 1997) of the total radiocesium deposited within a forest ecosystem. However, these estimates are based on the assumption radionuclide concentration in fungal sporocarps is similar to that of the fungal parts of mycorrhizae (Nikolova et al., 1997). The activity concentration in sporocarps is probably higher than in the mycelium (Vinichuk & Johanson, 2003, 2004) and sporocarps constitute only about 1% of the total mycelia biomass in a forest ecosystem. Due to the high levels of ^{137}Cs in sporocarps, their contribution to the internal dose in man may be high through consumption of edible mushrooms (Kalač, 2001). Consequently, the consumption of sporocarps of edible fungi (Skuterud et al., 1997) or of game animals that consumed large quantities of fungi with high ^{137}Cs contents (Johanson & Bergström, 1994) represents an important pathway by which ^{137}Cs enters the human food system.

The ^{137}Cs activity concentration in edible fungi species has not decreased over the last 20 years (*Suillus variegatus*) or significantly increased (*Cantharellus* spp.) (Mascanzoni, 2009; Rosén et al., 2011).

1.1.2 ^{137}Cs, [133]Cs and alkali metals in fungi

Although fungi are important for ^{137}Cs uptake and migration in forest systems and since the Chernobyl accident, fungal species may contain high concentrations of radiocesium, the reasons and mechanisms for the magnitude higher concentration of radiocesium in fungi than in plants remains unclear (Kuwahara et al., 1998; Bystrzejewska-Piotrowska & Bazala, 2008). In addition to radiocesium, fungi effectively accumulate potassium (K), rubidium (Rb) and stable cesium (^{133}Cs) (Gaso et al., 2000) and the concentrations of ^{137}Cs, ^{133}Cs and Rb in fungal sporocarps can be one order of magnitude higher than in plants growing in the same forest (Vinichuk et al., 2010b).

The chemical behavior of the alkali metals, K, Rb and ^{133}Cs, can be expected to be similar to ^{137}Cs, due to similarities in their physicochemical properties, e.g. valence and ion diameter (Enghag, 2000). Potassium is a macronutrient and an obligatory component of living cells, which depend on K+ uptake and K+ flux to grow and maintain life. In radioecology cesium is assumed to behave similarly to potassium. At the cellular level, K is accumulated within cells and is the most important ion for creating membrane potential and excitability. Myttenaere et al. (1993) summarize the relationship between radiocesium and K in forests and suggest the possible use of K as an analogue for predicting radiocesium behavior.

Generally, ^{137}Cs is positively associated with K concentration across plant species in an undisturbed forest ecosystem, which suggests ^{137}Cs, stable ^{133}Cs and K are assimilated in a similar way and the elements pass through the biological cycle together (Chao et al., 2008). Cs influx into cells and its use of K transporters is reviewed by White & Broadley (2000) and potassium transport in fungi is reviewed by Rodríguez-Navarro (2000).

Rubidium is another rarely studied alkali metal, which may be an essential trace element for organisms, including fungi. However, there is scarce information on the concentrations and distribution of Rb in fungi and its behavior in food webs originating in the forest. Rubidium is often used in studies on K uptake and appears to emulate K to a high degree (Marschner, 1995): both K and Rb have the same uptake kinetics and compete for transport along concentration gradients in different compartments of soil and organisms (Rodríguez-Navarro, 2000). The concentrations of K, Rb and ^{133}Cs have been analyzed in fungal sporocarps (Baeza et al., 2005; Vinichuk et al., 2010b; 2011) and a relation between the uptake of Cs and K has been found (Bystrzejewska-Piotrowska & Bazal, 2008). Cesium uptake in fungi is affected by the presence of K and Rb and the presence of ^{133}Cs (Gyuricza et al., 2010; Terada et al., 1998). Although in fungal sporocarps, the relationships between these alkali metals and ^{137}Cs when taken up by fungi and their underlying mechanisms are insufficiently understood, as Cs does not always have high correlation with K and it is suggested there is an alternative pathway for Cs uptake into fungal cells (Yoshida & Muramatsu, 1998).

The correlations between ^{137}Cs and these alkali metals suggest the mechanism of fungal uptake of ^{133}Cs and ^{137}Cs is different from K and that Rb has an intermediate behavior between K and ^{133}Cs (Yoshida & Muramatsu, 1998). However, this interpretation is based on a few sporocarp analyses from each species, and comprised different ectomycorrhizal and saprotrophic fungal species. Although fungal accumulation of ^{133}Cs is reported as species-dependent, there are few detailed studies of individual species (Gillet & Crout, 2000). The variation in ^{137}Cs levels within the same genotype of fungal sporocarps can be as large as the variation among different genotypes (Dahlberg et al., 1997).

Another way to interpret and understand the uptake and relations between ^{137}Cs, ^{133}Cs, K and Rb in fungi is to use the isotopic (atom) ratio ^{137}Cs/^{133}Cs. Chemically, ^{133}Cs and ^{137}Cs are the same, but the atom abundance and isotopic disequilibrium differ. Among other factors, uptake of ^{133}Cs and ^{137}Cs by fungi depends on whether equilibrium between the two isotopes is achieved. An attainment of equilibrium between stable ^{133}Cs and ^{137}Cs in the bioavailable fraction of soils within forest ecosystems is reported Karadeniz & Yaprak (2007) but in cultivated soils, equilibrium between fallout ^{137}Cs and stable [133]Cs among exchangeable, organic bound and strongly bound fractions has not reached, even though most ^{137}Cs was deposited on the soils more than 20 years before (Tsukada, 2006).

The important roles fungi play in nutrient uptake in forest soils, in particular its role in ^{137}Cs transfer between soil and fungi, requires better understanding of the mechanisms involved. Although transfer of radioactive cesium from soils to plants through fungi is well researched, there is still limited knowledge on natural stable [133]Cs and other alkali metals (K and Rb) and the potential role as a predictor for radiocesium behavior, and less is known about the relationships between [133]Cs and other alkali metals (K and Rb) during uptake by fungi.

To explore mechanisms governing the uptake of radionuclides (^{137}Cs) data on uptake of stable isotopes of alkali metals (K, Rb, [133]Cs) by fungal species, and the behavior of the three alkali metals K, Rb and [133]Cs in bulk soil, fungal mycelium and sporocarps are required. Therefore, an attempt was made to quantify the uptake and distribution of the alkali metals in the soil–mycelium–sporocarp compartments and to study the relationships between K, Rb and [133]Cs in the various transfer steps. Additionally, the sporocarps of ectomycorrhizal fungi *Suillus variegatus* were analyzed to determine whether i) Cs ([133]Cs and ^{137}Cs) uptake was correlated with K uptake; ii) intraspecific correlation of these alkali metals and ^{137}Cs activity concentrations in sporocarps was higher within, rather than among different fungal species; and, iii) the genotypic origin of sporocarps affected uptake and correlation.

Substantial research in this area has been conducted in Sweden after the fallout from nuclear weapons tests and the Chernobyl accident. Some results are published in a series of several articles in collaboration with Profs K.J. Johanson, H. Rydin and Dr. A. Taylor (Vinichuk et al., 2004; 2010a; 2010b; 2011).

This chapter aims to summarize the acquired knowledge from studies in Sweden and to place them in a larger context. The results are summarized and discussed and address the issues of K, Rb and [133]Cs concentrations in soil fractions and fungal compartments (Section 1.3.); concentration ratios of K, Rb and [133]Cs in soil fractions and fungi (Section 1.4); relationships between K, Rb and [133]Cs in soil and fungi (Section 1.5); the isotopic (atom) ratios ^{137}Cs/K, ^{137}Cs/Rb and ^{137}Cs/[133]Cs in fungal species (Section 1.6); K, Rb and Cs (^{137}Cs and [133]Cs) in sporocarps of a single species (Section 1.7); mechanisms of ^{137}Cs and alkali metal uptake by fungi (Section 1.8); Cs (^{137}Cs and [133]Cs), K and Rb in *Sphagnum* plants (Section 2); distribution of Cs (^{137}Cs and [133]Cs), K and Rb within *Sphagnum* plants (Section 2.3); mass concentration and isotopic (atom) ratios between ^{137}Cs, K, Rb and [133]Cs in segments of *Sphagnum* plants (Section 2.4); relationships between ^{137}Cs, K, Rb and [133]Cs, in segments of *Sphagnum* plants (Section 2.5); mechanisms of ^{137}Cs and alkali metal uptake by *Sphagnum* plants (Section 2.6); and conclusions from the Swedish studies (Section 3). Before presenting and discussing results a short description of study area, study design and methods used is presented (section 1.2).

1.2 Study area, study design and methods for results presented
1.2.1 Study area

The K, Rb and [133]Cs concentrations in soil fractions and fungal compartments were studied in an area located in a forest ecosystem on the east coast of central Sweden (60°22′N, 18°13′E). The soil was a sandy or clayey till and the humus mainly occurred in the form of mull. A more detailed description of the study area is presented by Vinichuk et al. (2010b).

Sporocarps of ectomycorrhizal fungi *Suillus variegatus* was studied in an area located about 40 km north-west of Uppsala in central Sweden (N 60°08'; E 17°10'). The forest is located on moraine and is dominated by Scots pine (*Pinus sylvestris*) and Norway spruce (*Picea abies*), with inserts of deciduous trees, primarily birch (*Betula pendula* and *Betula pubescens*). The field layer consisted mainly of the dwarf shrubs bilberry (*Vaccinium myrtillus* L.), lingonberry (*Vaccinium vitis-idaea* L.) and heather *Calluna vulgaris* L.): for details about the area and sampling see Dahlberg et al. (1997).

1.2.2 Study design

For studies of K, Rb and ^{133}Cs concentrations in soil fractions and fungal compartments, samples of soil and fungal sporocarps were collected from 10 sampling plots during September to November 2003. Four replicate soil samples were taken, with a cylindrical steel tube with a diameter of 5.7 cm, from around and directly underneath the fungal sporocarps (an area of about 0.5 m²) and within each 10 m² area to a depth of 10 cm. Soil cores were divided horizontally into two 5-cm thick layers. Sporocarps of 12 different fungal species were collected and identified to species level, and the ^{137}Cs activity concentration in fresh material was determined. The sporocarps were dried at 35°C to constant weight and concentrations of ^{133}Cs, K and Rb were determined.

A selection of dried sporocarps of *S. variegatus* (n=51), retained from a study by Dahlberg et al. (1997) on the relationship between ^{137}Cs activity concentrations and genotype identification, was used. The sporocarps were collected once a week during sporocarp season (end of August through September) in 1994 and were taken from five sampling sites (100 to 1600 m² in size) within an area of about 1 km². Eight genotypes with 2 to 8 sporocarps each were tested (in total 32 sporocarps) and are referred to here as individual genotypes. Sporocarps within genotypes were spatially separated by up to 10-12m. All genotypes were used for the estimation of correlation coefficients, but only genotypes with at least four sporocarps were included in the alkali metal analysis. In addition, 19 individual sporocarps with unknown genotype (i.e. not tested for genotype identity) were included: these sporocarps consisted of both the same and different genotypes. The combined set of sporocarps refers to all sporocarps: for further details about the sampling and identification of genotypes see Dahlberg et al. (1997). The ^{137}Cs activity concentration values corrected to sampling date and expressed as kBq kg^{-1} dry weight (DW) for each sporocarp, as reported by Dahlberg et al. (1997), were used.

1.2.3 Methods

For the studies of K, Rb and ^{133}Cs concentrations in soil fractions and fungal compartments, fungal mycelia were separated from the soil samples (30–50 g, 0–5 cm layer depth) under a dissection microscope (magnification X64) with forceps and by adding small amounts of distilled water to disperse the soil. The prepared fraction of mycelium (30–60 mg DW g^{-1} soil) was not identified to determine of the mycelia extracted from the soil samples and the sporocarps belonged to the same species, as it assumed a majority of the prepared mycelia belonged to the same species as the nearby sporocarps. The method for mycelium preparation is described in Vinichuk & Johanson (2003). Mycelium samples were dried at 35°C to constant weight for determination of K, Rb and ^{133}Cs.

The soil samples (0–5 cm layer) were partitioned by the method described in Gorban & Clegg (1996). First, soil was gently sieved through a 2 mm mesh giving a bulk soil fraction. The remaining soil aggregates containing roots were further crumbled and gently squeezed between the fingers: this was called the rhizosphere fraction. The residue (finest roots with adhering soil particles) was called the soil–root interface fraction. Nine samples of bulk soil fraction and mycelium, 12 samples of fungal sporocarps, and six samples of rhizosphere and soil–root interface fraction were analyzed for K, Rb and [133]Cs.

The ^{137}Cs activity concentrations in the bulk soil samples and sporocarps were determined with calibrated HP-Ge detectors, corrected to sampling date and expressed as Bq kg[-1] DW. The measuring time employed provided a statistical error ranging between 5 and 10%. For element analyses, a 2.5 g portion of each sample was analyzed by inductively coupled plasma in the laboratories of ALS Scandinavia (Luleå, Sweden) with recoveries 97–101% for K; 97.5–99.4% for Rb, and 93.7–102.5% for, [133]Cs. For soil, CRM SO-2 (heavy metals in soil) was used which had no certified values for K, Rb or [133]Cs. Element concentrations in the analyzed fractions are reported as mg kg[-1] DW.

For element analyses (K, Rb and [133]Cs) of S. variegatus sporocarps, aliquots of about 0.3 g of each sample were analyzed by the same technique. Element concentrations are reported as mg kg[-1] DW and the isotopic ratio of ^{137}Cs/[133]Cs was calculated with Equations 1 and 2 (Chao et al., 2008):

$$\frac{^{137}Cs}{^{133}Cs} = \frac{A}{C} \times \frac{\alpha}{\lambda \times N} \times 10^3 \tag{1}$$

where: A is the ^{137}Cs radioactivity (Bq kg[-1]); λ is the disintegration rate of [137]Cs 7.25 x10[-10] s[-1]; a is the atomic weight of cesium (132.9); N is the Avogadro number, which is 6.02 x10[23]; and, [133]C and C are the [133]Cs concentration (mg g[-1]). Eq. (1) can be simplified to Eq. 2:

$$\frac{^{137}Cs}{^{133}Cs} = 3.05 \times 10^{-10} \times \frac{A}{C} \tag{2}$$

where: A is the ^{137}Cs activity concentration in Bq kg[-1] and [133]C is the [133]Cs concentration in mg kg[-1]. Thus, the units of the isotope ratio are dimensionless.

Relationships between K, Rb, [133]Cs and ^{137}Cs concentrations in different fractions and sporocarps of S. variegatus were identified by Pearson correlation coefficients. Correlation coefficients were analyzed in five separate sets of samples: in four sets, all samples had known genotype identity, and in the last set, there was a combined set of samples containing both genotypes that had been tested by somatic incompatibility sporocarps and genotypes that had not been tested. Correlation analyses for genotypes with three or less sporocarps were omitted. All statistical analyses were run with Minitab® 15.1.1.0. (© 2007 Minitab Inc.) software, with level of significance of 5% (0.05), 1% (0.01) and 0.1% (0.001).

1.3 K, Rb and [133]Cs concentrations in soil fractions and fungal compartments

K, Rb and [133]Cs concentrations values in soil fractions and fungal compartments are necessary for calculating the concentration ratio at each step of its transfer in the soil-fungi system, differences in the uptake between elements and the relationships. This in turn will be the main reason for the different K, Rb and [133]Cs concentrations observed in sporocarps of various fungal species. Concentrations of K, Rb and [133]Cs in bulk soil were not significantly different from those in the rhizosphere, although the values for all three elements were slightly higher in the rhizosphere fraction (Table 1).

Element	Bulk soil	Rhizosphere	Soil root-interface	Fungal mycelium	Fruit bodies
K	642.6 (214.6)a	899.3 (301.4)a	3215 (842.8)b	2 867(727.5)b	43 415 (20 436)b
Rb	3.9 (2.7)a	5.4 (4.4)a	6.8 (1.7)a	13.8 (6.9)b	253.9 (273.6)b
^{133}Cs	0.3 (0.2)a	0.4 (0.3)a	0.2 (0.05)a	0.8 (0.8) a	5.65 (7.1)b

[1]Means within rows with different letters (a or b) are significantly different (p < 0.001).

Table 1. Mean concentrations of K, Rb and ^{133}Cs (mg kg^{-1} DW (standard deviation)) in soil fractions and fungi[1].

Potassium concentrations were higher in both the soil–root interface and fungal mycelium fractions than in the bulk soil and rhizosphere fraction. A comparison of K, Rb and ^{133}Cs concentrations revealed fungal sporocarps accumulated much greater amounts of these elements than mycelium. For example, K concentrations in fungal sporocarps collected from the same plots where soil samples and mycelium were extracted were about 15 times higher than K concentrations found in mycelium. The concentrations of Rb in fungal sporocarps were about 18-fold higher than in corresponding fungal mycelium, and those of ^{133}Cs were about 7-fold higher (Table 1).

Thus, potassium concentration increased in the order bulk soil<rhizosphere<fungal mycelium<soil–root interface<fungal sporocarps and was higher in the soil–root interface fraction and fungi than in bulk soil. The high concentrations of K in fungal sporocarps may reflect a demand for this element as a major cation in osmoregulation and that K is an important element in regulating the productivity of sporophore formation in fungi (Tyler, 1982).

Rb in mycelium was 3.5-fold higher than in bulk soil and 2.5-fold higher than in rhizosphere, and concentrations increased in the order bulk soil<rhizosphere<soil–root interface<fungal mycelium<fungal sporocarps. The concentrations of Rb were slightly higher in the soil–root interface fraction than in bulk soil; thus, fungi appeared to have high preference for this element, as the accumulation of Rb by fungi, and especially fungal sporocarps, was pronounced. Rubidium concentrations in sporocarps were more than one order of magnitude higher than those in mycelium extracted from soil of the same plots where fungal sporocarps were sampled. The ability of fungi to accumulate Rb is documented: mushrooms accumulate at least one order of magnitude higher concentrations of Rb than plants growing in the same forest (Yoshida & Muramatsu, 1998).

Concentrations of stable cesium varied considerably among samples but no significant differences were found among the different fractions analyzed. Cesium concentrations increased in the order soil–root interface<bulk soil<rhizosphere<fungal mycelium<fungal sporocarps, and were only significantly higher in fungal sporocarps, compared with bulk soil. Stable ^{133}Cs was generally evenly distributed within bulk soil, rhizosphere and soil–root interface fractions, indicating no ^{133}Cs enrichment in those forest compartments. However, ^{133}Cs concentrations in sporocarps were nearly one order of magnitude higher than those found in soil mycelium.

Radioactive ^{137}Cs presented similar to ^{133}Cs behavior, where ^{137}Cs activity increased in the order soil<mycelium<fungal sporocarps (Vinichuk & Johanson, 2003; Vinichuk et al., 2004). The differences between fungal species in their preferences for uptake of ^{137}Cs or stable ^{133}Cs appear to reflect the location of the fungal mycelium relative to that of cesium within the soil profile (Rühm et al., 1997). Unlike ^{137}Cs, stable ^{133}Cs originates from soil; therefore, the

amount of unavailable ^{133}Cs, compared to the total amount of ^{133}Cs, in soil presumably higher than that of ^{137}Cs. As a result, stable ^{133}Cs is considered less available for uptake as it is contained in mineral compounds and is difficult for fungi or plants to access: the concentration ratio of stable ^{133}Cs in mushrooms is lower than for ^{137}Cs (Yoshida & Muramatsu, 1998). The differing behavior of the natural and radioactive forms of ^{133}Cs may derive from their disequilibrium in the ecosystem (Horyna & Řanad, 1988).

1.4 Concentration ratios of K, Rb and ^{133}Cs in soil fractions and fungi

The concept of concentration ratios (CR, defined as concentration of the element (mg kg^{-1} DW) in a specific fraction or fungi divided by concentration of the element (mg kg^{-1} DW) in bulk soil) is widely used to quantify the transfer of radionuclides from soil to plants/fungi. This approach allows the estimation of differences in uptake of elements. The elements concentration ratio data followed a similar pattern, but the enrichment of all three elements in fungal material was more evident, particularly in the sporocarps (Table 2).

Element	Rhizosphere	Soil root-interface	Fungal mycelium	Fruit bodies
K	1.7 (0.4)	6.1 (1.9)	5.1 (1.4)	68.9 (23.1)
Rb	1.3 (0.4)	2.7 (1.1)	3.9 (1.1)	121.7 (172.2)
Cs	1.1 (0.5)	0.8 (0.3)	2.1 (0.9)	39.7 (67.6)

Table 2. Concentration ratios CR (defined as concentration of the element (mg kg^{-1} DW) in the specific fraction divided by concentration of the element (mg kg^{-1} DW) in bulk soil) (mean values (standard deviation)).

Thus, for all three alkali metals studied, the levels of K, Rb, ^{133}Cs and ^{137}Cs in sporocarps were at least one order of magnitude higher than those in fungal mycelium (Table 2). The concentration ratios for each element varied considerably between the species sampled. The saprotrophic fungus *Hypholoma capnoides* had the lowest values and the mycorrhizal fungus *Sarcodon imbricatus* had the highest. Sporocarp:bulk soil concentration ratios are presented in Table 3.

Sarcodon imbricatus accumulates nearly 100 000 Bq kg^{-1} of ^{137}Cs, giving TF values (defined as ^{137}Cs activity concentration (Bq kg^{-1} DW) in fungi divided by ^{137}Cs deposition (kBq m^{-2})) about 22 (Vinichuk & Johanson, 2003). The sporocarps of *Sarcodon imbricatus* had distinctively high concentration ratios of Rb and ^{133}Cs than other species analyzed. The mycorrhizal fungus *Cantharellus tubaeformis*, is another species showing relatively high concentration ratios, particularly for K and Rb. *Cantharellus tubaeformis* accumulates several tens of thousands Bq kg^{-1} of ^{137}Cs (Kammerer et al., 1994). Among those with moderate concentration ratios for each element are *Boletus edulis*, *Tricholoma equestre*, *Lactarius scrobiculatus* and *Cortinarius* spp.

Thus, the levels of K, Rb, ^{133}Cs and ^{137}Cs in sporocarps were at least one order of magnitude higher than those in fungal mycelium indicating biomagnification through the food web in forest ecosystems.

Plot	Species	Concentration ratios		
		K	Rb	^{133}Cs
4	Boletus edulis	62.7	77.4	37.4
6	Cantharellus tubaeformis	104.7	109.7	15.5
7	Cortinarius armeniacus	67.5	69.6	19.2
5	C. odorifer	71.8	70.9	34.7
8	C. spp.	90.9	157.2	14.8
8-10	Hypholoma capnoides[1]	26.6	13.1	6.9
1	Lactarius deterrimus	29.9	17.2	2.6
3	L. scrobiculatus	67.8	26.2	3.7
6	L. trivialis	77.5	126.9	52.2
5-7	Sarcodon imbricatus	101.7	675.7	258.8
2	Suillus granulates	58.6	41.4	14.7
10-11	Tricholoma equestre	66.6	75.4	15.4

[1]Saprophyte, all other analyzed fungal species are ectomycorrhizal

Table 3. Element concentration ratios (mg kg^{-1} DW in fungi)/(mg kg^{-1} DW in bulk soil) in fungi for fungal sporocarps.

1.5 Relationships between K, Rb and ^{133}Cs in soil and fungi

Although correlation analysis may be not definitive, it is a useful approach for elucidating similarities or differences in uptake mechanisms of cesium (^{137}Cs and ^{133}Cs), K and Rb: close correlation between elements indicates similarities in their uptake mechanisms. No significant correlations between K in soil and in either mycelium ($r=0.452$, ns) or in sporocarps ($r=0.338$, ns) has been identified and sporocarp Rb and ^{133}Cs concentrations were unrelated to soil concentrations, however, in mycelium both elements were correlated with soil concentrations (Rb: $r=0.856$, $p=0.003$; Cs: $r=0.804$, $p=0.009$). There was a close positive correlation ($r=0.946$, $p=0.001$) between the K:Rb ratio in soil and in fungal mycelium (Figure 1b) and this relationship was also apparent between soil and sporocarps, but was weak and not significant ($r=0.602$, ns: Figure1b).

The K:^{133}Cs ratio in soil and fungal components had a different pattern: the K:Cs ratio in mycelium was closely positively correlated ($r=0.883$, $p=0.01$) to the K:^{133}Cs ratio in soil (Figure 1a), but was relatively weakly and non-significantly correlated to soil in fungal sporocarps. No significant correlations were found between the concentrations of the three elements in fungi, soil pH or soil organic matter content (data not shown).

The competition between K, Rb and ^{133}Cs in the various transfer steps was investigated in an attempt to estimate the relationships between the concentrations of these three elements in soil, mycelia and fungal sporocarps. The lack of a significant correlation between K in soil and in either mycelium or sporocarps indicated a demand for essential K in fungi, regardless of the concentration of this element in soil. Regardless of fungal species, K concentration in fungi appears to be controlled within a narrow range, (Yoshida & Muramatsu, 1998), and supports the claim K uptake by fungi is self-regulated by the internal nutritional requirements of the fungus (Baeza et al., 2004).

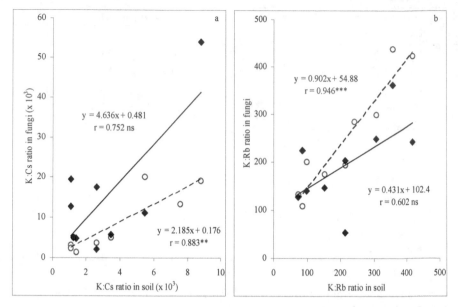

Fig. 1. Ratio of (a) K: [133]Cs and (b) K:Rb in fungal sporocarps (♦, solid line) and soil mycelium (○, dotted line) in relation to the soil in which they were growing. ** p=0.01, *** p=0.001

The relationships observed between K:Rb and K:[133]Cs ratios in fungal sporocarps and soil mycelia, with respect to the soil in which they were growing (Figure 1), also indicated differences in uptake of these alkali metals by fungi. Although correlation analyses is not the best tool for analyzing the uptake mechanism, the closest positive correlations between K:Rb ratios in fungal mycelium and in soil indicated similarities in the uptake mechanism of these two elements by fungi, although the relationships between K: [133]Cs ratios in soil mycelium and in soil were less pronounced. These findings were in good agreement with the suggestion by Yoshida & Muramatsu (1998) that there might be an alternative pathway for [133]Cs uptake into cells and the mechanism of [133]Cs uptake by fungi could be similar to that for Rb, as [133]Cs does not show a good correlation with K. The high efficiency of Rb uptake by fungi indicates Rb, but not [133]Cs, eventually replaces essential K due to K limitation (Brown & Cummings, 2001) and Rb has the capacity to partially replace K, but [133]Cs does not (Wallace, 1970 and references therein). Forest plants apparently discriminate between K+ and Rb+ in soils and a shortage of K+ favors the uptake of the closely related Rb+ ion (Nyholm & Tyler, 2000), whereas, increasing K+ availability in the system decreases Rb+ uptake (Drobner & Tyler, 1998). These results provided new insights into the use of transfer factors or concentration ratios.

1.6 The isotopic (atom) ratios ^{137}Cs/K, ^{137}Cs/Rb and ^{137}Cs/[133]Cs in fungal species

The isotopic ratios of ^{137}Cs/K, ^{137}Cs/Rb and ^{137}Cs/[133]Cs in the fungal sporocarps belonging to different species were used to interpret the distribution of ^{137}Cs and the alkali metals in fungi and to provide better understanding of its uptake mechanisms. Measurements of trace levels of stable [133]Cs could be another way of obtaining information about the biological behavior of [137]Cs. To obtain better estimates, the isotopic ratios for fungal sporocarps in this study (Vinichuk et al., 2010b) were calculated and compared with estimates calculated in similar studies by Yoshida & Muramatsu (1998). Mean values of isotopic ratios of ^{137}Cs/K, ^{137}Cs/Rb and ^{137}Cs/[133]Cs in the fungal sporocarps, and range and correlation coefficients between concentration ratios ^{137}Cs/[133]Cs and K, Rb and [133]Cs are presented in Table 4.

Data set	n	Isotopic ratios		
		^{137}Cs/K	^{137}Cs/Rb	^{137}Cs/^{133}Cs
Vinichuk et al. (2010b), Sweden	12	$14.4(1.54–45.4)\times10^{-13}$	$7.8(0.55–30.9)\times10^{-10}$	$4.9(0.30–15.1)\times10^{-8}$
Yoshida & Muramatsu (1998), Japan	29	$5.2(0.15–23.0)\times10^{-16}$	$3.4(0.14–18.2)\times10^{-13}$	$4.1(1.53–5.94)\times10^{-9}$
		Correlation coefficients		
		^{137}Cs/^{133}Cs:K	^{137}Cs/^{133}Cs:Rb	^{137}Cs/^{133}Cs:^{133}Cs
Vinichuk et al. (2010b), Sweden	12	0.25	−0.35	−0.31
Yoshida & Muramatsu (1998), Japan	29	0.12	0.39	0.26

Table 4. Isotopic (atom) ratios of ^{137}Cs/K, ^{137}Cs/Rb, ^{137}Cs/[133]Cs, correlation coefficients between isotopic ratios ^{137}Cs/[133]Cs and mass concentrations of K, Rb and [133]Cs in fungal sporocarps (n = number of sporocarps analyzed).

The activity concentrations of ^{137}Cs in fungal sporocarps were about 13 to 16 orders of magnitude lower than mass concentrations of K, 10 to 13 orders of magnitude lower than mass concentrations for Rb, and 8 to 9 orders of magnitude lower than mass concentrations for [133]Cs. Isotopic (atom) ratios in the fungal sporocarps collected in Sweden were two-three orders of magnitude narrower than those collected in Japan, which reflected the level of ^{137}Cs concentrations in mushrooms: the median value for all fungi species was 4151 Bq kg^{-1} DW in Swedish forests and 135 Bq kg^{-1} DW in Japanese forests. Isotopic (atom) ratios of ^{137}Cs/K, ^{137}Cs/Rb, ^{137}Cs/[133]Cs were variable in both datasets and appeared independent of specific species of fungi. These ratios might reflect the isotopic ratios of soil horizons from which radiocesium is predominantly taken up and be a possible source of the variability in isotopic ratios in fungal fruit bodies. Rühm et al. (1997) used the isotopic ratio [134]Cs/^{137}Cs to localize mycelia of fungal species *in situ*; alternatively, the isotopic (atom) ratio ^{137}Cs/[133]Cs can be used to localize fungal mycelia *in situ*. However, this approach is only appropriate for organic soil layers, which contain virtually no or very little clay mineral to which cesium can bind. The isotopic ratios ^{137}Cs/[133]Cs in fruit bodies of fungi were similar to those found in organic soil layers of forest soil (Rühm et al., 1997; Karadeniz & Yaprak, 2007).

The relationships observed between the concentration ratios ^{137}Cs/[133]Cs and K, Rb and [133]Cs in fungal sporocarps also varied widely and were inconsistent (Table 4). The concentration of K, Rb and [133]Cs in sporocarps appeared independent of the ^{137}Cs/[133]Cs isotopic ratio, suggesting differences in uptake of these alkali metals by fungi and complex interactions between fungi, their host and the environment.

1.7 K, Rb and Cs (^{137}Cs and ^{133}Cs) in sporocarps of a single species

Most results presented in this Chapter are already published (Vinichuk et al., 2011), and are based on sporocarp analysis of different ectomycorrhizal and saprotrophic fungal species. Fungal accumulation of ^{137}Cs is suggested to be species-dependent, thus, ^{137}Cs activity concentration and mass concentration of K, Rb and ^{133}Cs in fungal sporocarps belonging to the mycorrhizal fungus *Suillus variegatus* were analyzed. *S. variegatus* form *mycorrhiza* with Scots pine and predominantly occur in sandy, acidic soils and have a marked ability to accumulate radiocesium (Dahlberg et al., 1997): as this is an edible mushroom, high radiocesium contents present some concern with regard to human consumption.

The concentrations of K (range 22.2-52.1 g kg^{-1}) and Rb (range 0.22-0.65 g kg^{-1}) in sporocarps of *S. variegatus* varied in relatively narrow ranges, whereas, the mass concentration of ^{133}Cs had a range of 2.16 to 21.5 mg kg^{-1} and the activity concentration of ^{137}Cs ranged from 15.8 to 150.9 kBq kg^{-1}. Both ^{133}Cs and ^{137}Cs had wider ranges than K or Rb within sporocarps from the same genotype or across the combined set of sporocarps (Table 5). The mean of the ^{137}Cs/^{133}Cs isotopic ratio in the combined set of sporocarps was 2.5×10^{-7} (range 8.3×10^{-8} and 4.4×10^{-7}). The ^{137}Cs/Cs isotopic ratios from identified genotypes were site-genotype dependent: the ratio values of genotypes at site 4 were about two-times higher than the ratios of genotypes at site 2 (Table 6).

Site-genotype[1]	n	K			Rb			^{133}Cs			^{137}Cs		
		g kg^{-1}	%		g kg^{-1}	%		mg kg^{-1}	%		kBq kg^{-1}	%	
		M	SD	CV	M	SD	CV	M	SD	CV	M	SD	CV
Sporocarps with identified genotypes													
2-1	8	30.6	8.06	26.4	0.47	0.12	24.7	12.1	4.23	35.1	67.3	35.1	52.2
2-2	6	28.0	6.99	25.0	0.50	0.07	13.8	16.6	2.19	13.2	75.9	23.2	30.6
4-3	4	28.5	2.13	7.5	0.39	0.16	4.0	6.6	0.44	6.7	68.9	11.7	17.0
4-4	3	33.6	8.60	-	0.30	0.04	-	3.0	0.60	-	39.1	9.38	-
4-5	2	38.9	2.40	-	0.36	0.02	-	3.8	0.04	-	35.7	28.2	-
4-6	2	35.2	8.84	-	0.37	0.11	-	3.7	2.16	-	26.8	8.54	-
7-7	5	33.7	5.79	17.2	0.34	0.06	17.9	6.7	0.80	12.0	71.4	9.30	13.0
6-8	2	25.4	1.34	-	0.31	0.03	-	8.7	2.16	-	63.3	18.3	-
Sporocarps with unknown genotypes													
	19	33.4	6.69	20.0	0.38	0.08	20.3	7.7	1.97	25.5	66.0	21.3	32.3
Combined set of sporocarps (identified and unknown genotypes)													
	51	31.9	6.79	21.3	0.40	0.09	23.6	8.7	4.36	50.1	63.7	24.2	38.0

[1]Site numbering according to Dahlberg et al. (1997), the second figure is a running number of the study's different genotypes.

Table 5. Potassium, rubidium and cesium (^{133}Cs) mass concentrations and ^{137}Cs activity concentrations in sporocarps of *S. variegatus* (DW) from identified and unknown genotypes, where n = number of sporocarps of each genotype analyzed, M = mean, SD = standard deviation, CV = coefficient of variation.

Site-genotype[1]	Identified genotypes								Unidentified genotypes	Combined set of sporocarps
	2-1	2-2	4-3	4-4	4-5	4-6	7-7	6-8		
M	1.67	1.43	3.16	3.95	2.86	2.43	3.27	2.24	2.62	2.50
CV (%)	97.1	36.4	10.4	5.1	78.1	29.5	9.2	3.9	20.0	34.6

[1]Site numbering according to Dahlberg et al. (1997), the second figure is a running number of the study's different genotypes.

Table 6. ^{137}Cs/^{133}Cs isotopic (atom) ratios in sporocarps of *S. variegatus* from identified genotypes, with unknown genetic belonging, and the two combined groupings, x10^{-7}. M = mean, CV = coefficient of variation.

Similarly, in results obtained from a previous study (Vinichuk et al. 2004) the concentrations of K in sporocarps of *S. variegatus* were not related to the concentrations of ^{137}Cs (r=0.103) or ^{133}Cs (r=−0.066) in the combined data set (Figure 2: c, b). In contrast, the concentrations of K and Rb were significantly correlated in the combined dataset (r=0.505, Figure 2: a).

Rubidium was strongly correlated with stable ^{133}Cs (r=0.746) and moderately correlated with ^{137}Cs (r=0.440) and K (r=0.505: Figure 2: d, e, a). Both ^{133}Cs and ^{137}Cs were significantly correlated in the combined dataset (Figure 2: f).

The ^{137}Cs/^{133}Cs isotopic ratio in the combined dataset was not correlated to K concentration, but correlated moderately and negatively with both ^{133}Cs (r=−0.636) and Rb (r=−0.500) concentrations (Figure 3: a, c, b).

Thus, the study of *S. variegatus* revealed no significant correlations between ^{133}Cs mass concentration or ^{137}Cs activity concentration and the concentration of K in sporocarps, either within the whole population or among the genotypes.

Potassium, ^{133}Cs and ^{137}Cs within the four genotypes were also not correlated, with one genotype exception (Table 7). However, the exception was conditional due to a one single value. Three of four analyzed sporocarp genotypes had high correlation between K and Rb: the forth was only moderately correlated (Table 7).

However, the correlations between ^{137}Cs and K and Rb and ^{133}Cs in the four genotypes were inconsistent (Table 3). Potassium, Rb, ^{133}Cs and ^{137}Cs were correlated in genotype 2-1 (due to one single value), whereas, no or negative correlations were found between the same elements/isotopes for the other three genotypes. In two of four genotypes, the ^{137}Cs/^{133}Cs isotopic ratio was not correlated with ^{133}Cs, K or Rb; however, there was a negative correlation with Rb in one genotype (2-2) and positive correlation with ^{133}Cs in another (4-3) (Table 7).

Data obtained for *S. variegatus* supported results from earlier studies (Ismail, 1994; Yoshida & Muramatsu, 1998) on different species of fungi, suggesting cesium (^{137}Cs and ^{133}Cs) and K are not correlated in mushrooms. Thus, correlation analysis may be a useful, although not definitive, approach for elucidating similarities or differences in uptake mechanisms of cesium (^{137}Cs and ^{133}Cs) and K.

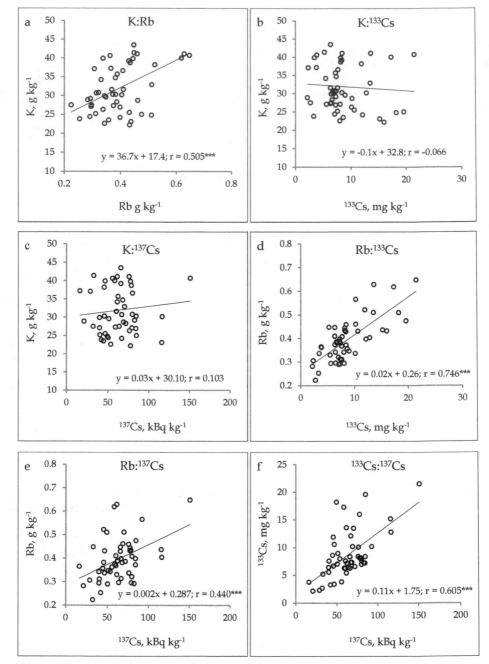

Fig. 2. Relationship between ^{137}Cs and K, Rb and ^{133}Cs concentrations in sporocarps in the combined set of all *S. variegatus* sporocarps (a-f). K:Rb (a); K:^{133}Cs (b); K:^{137}Cs (c); Rb:^{133}Cs (d); Rb:^{137}Cs (e); and, ^{133}Cs:^{137}Cs (f). *** p=0.001

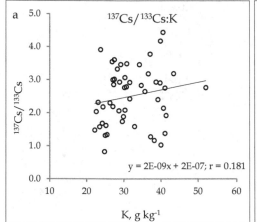

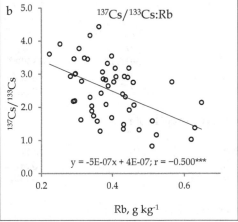

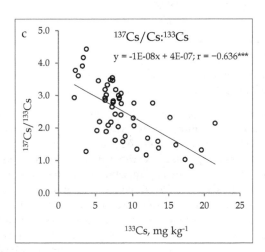

Fig. 3. Relationship between the ^{137}Cs/^{133}Cs isotopic (atom) ratios ($\times 10^{-7}$) and K, Rb and ^{133}Cs mass concentrations in the combined set of *S. variegatus* sporocarps, (a) ^{137}Cs/^{133}Cs:K; (b) ^{137}Cs/^{133}Cs:Rb; and, (c) ^{137}Cs/^{133}Cs:^{133}Cs. *** p=0.001

	^{137}Cs	K	Rb	^{133}Cs
Genotype 2-1 (8 sporocarps)				
K	0.502			
Rb	0.626*	0.966***		
^{133}Cs	0.908**	0.745*	0.837**	
^{137}Cs/^{133}Cs		−0.172	−0.058	0.240
Genotype 2-2 (6 sporocarps)				
K	−0.472			
Rb	−0.658	0.928**		
^{133}Cs	−0.263	−0.138	0.159	
^{137}Cs/^{133}Cs		−0.352	−0.608	−0.586
Genotype 4-3 (4 sporocarps)				
K	−0.531			
Rb	0.177	0.696		
^{133}Cs	0.979*	−0.569	0.182	
^{137}Cs/^{133}Cs		−0.488	0.163	0.930
Genotype 7-7(5 sporocarps)				
K	−0.562			
Rb	−0.472	0.987**		
^{133}Cs	0.699	−0.528	−0.404	
^{137}Cs/^{133}Cs		−0.115	−0.155	−0.345

[1]* p=0.05; ** p=0.01; *** p=0.001

Table 7. Correlation coefficients between concentrations of potassium, rubidium and cesium (^{133}Cs and ^{137}Cs) in genotypes of *S. variegatus* with more than four sporocarps analyzed[1].

The concentration of K in sporocarps appeared independent of the ^{137}Cs/^{133}Cs isotopic ratio in both the whole population (Figure 3) and among the genotypes, with one exception (Table 7). The absence of correlation between ^{137}C (or ^{133}Cs) and K in fungi may be due to the incorporation of K being self-regulated by the nutritional requirements of the fungus, whereas, incorporation of ^{137}Cs is not self-regulated by the fungus (Baeza et al., 2004). Although K and cesium (^{133}Cs and ^{137}Cs) concentrations did not correlate within *S. variegatus*, both K$^+$ and Cs$^+$ ions may compete for uptake by fungi. In experiments under controlled conditions and with sterile medium (Bystrzejewska-Piotrowska & Bazala, 2008), the competition between Cs$^+$ and K$^+$ depends on Cs$^+$ concentration in the growth medium and on the path of Cs$^+$ uptake. In studies of Cs uptake by hyphae of basidiomycete *Hebeloma vinosophyllum* when grown on a simulated medium (Ban-Nai et al., 2005), the addition of monovalent cations of K$^+$, Rb$^+$, and NH$_4$$^+$ reduced uptake of Cs. In addition, radiocesium transport by arbuscular mycorrhizal (AM) fungi decreases if K concentration increases in a compartment accessible only to AM (Gyuricza et al., 2010), and a higher Cs:K ratio in the nutrient solution increases uptake of Cs by ectomycorrhizal seedlings (Brunner et al., 1996). A noticeable (20-60%) and long-lasting (at least 17 years) reduction in ^{133}Cs activity concentration in fungal sporocarps *in situ* due to a single K fertilization of 100 kg ha^{-1} in a Scots pine forest is reported by Rosén et al., (2011).

The relation between ^{137}Cs and K, and Rb and ^{133}Cs within S. *variegatus* (Figure 2) was similar to an earlier report on different species of fungi (Yoshida & Muramatsu, 1998). Rubidium concentration in sporocarps was positively correlated with ^{133}Cs and ^{137}Cs, but generally negatively correlated with ^{137}Cs/^{133}Cs isotopic ratio, i.e. a narrower ^{137}Cs/^{133}Cs ratio in sporocarps resulted in higher Rb uptake by fungi. This ratio may reflect the soil layers explored by the mycelia (Rühm et al., 1997), as fungi have a higher affinity for Rb than for K and cesium (Ban-Nai et al., 2005; Yoshida & Muramatsu, 1998), and Rb concentrations in sporocarps can be more than one order of magnitude greater than in mycelium extracted as fungal sporocarps from soil of the same plots (Vinichuk et al., 2011). Soil mycelia always consist of numerous fungal species and the intraspecific relationships between soil mycelia and sporocarps has not yet been estimated; however, the development of molecular methods with the ability to mass sequence environmental samples in combination with quantitative PCR may now enable such analysis to be conducted.

Mass concentration of ^{133}Cs and activity concentration of ^{137}Cs have different relations in fungal sporocarps: in three of four genotypes, there was a high correlation, two of which were significant (r=0.908** and r=979*), and there was no correlation in the fourth genotype (r=−0.263, Table 7), whereas, correlation between ^{137}Cs and ^{133}Cs within the whole population was only moderate (r=0.605*** Figure 2). In terms of ^{133}Cs and ^{137}Cs behavior, there would be no biochemical differentiation, but there could be differences in atom abundance and isotopic disequilibrium within the system. Fungi have large spatiotemporal variation in ^{133}Cs and ^{137}Cs content in sporocarps of the same species and different species (de Meijer et al., 1988), and the variation in K, Rb, ^{133}Cs and ^{137}Cs concentrations within a single genotype appeared similar, or lower, than the variation within all genotypes. The results for ^{137}Cs and alkali elements in a set of samples of S. *variegatus*, collected during the same season and consisting of sporocarps from both different and the same genotype, indicated the variability in concentrations was similar to different fungal species collected in Japan over three years (Yoshida & Muramatsu, 1998).

The relatively narrow range in K and Rb variation and the higher ^{133}Cs and ^{137}Cs variations might be due to different mechanisms being involved. The differences in correlation coefficients between ^{137}Cs and the alkali metals varied among and within the genotypes of S. *variegatus*, suggesting both interspecific and intrapopulation variation in the uptake of K, Rb, stable ^{133}Cs and, ^{137}Cs and, their relationships could be explained by factors other than genotype identity. The variability in ^{137}Cs transfer depends on the sampling location of fungal sporocarps (Gillett & Crout, 2000), for S. *variegatus*, these interaction factors might include the spatial pattern of soil chemical parameters, heterogeneity of ^{137}Cs fallout, mycelia location, and heterogeneity due to abiotic and biotic interactions increasing over time (Dahlberg et al., 1997).

Within the combined set of sporocarps the concentration of Rb and ^{137}Cs activity concentration in S. *variegatus* sporocarps were normally distributed but the frequency distribution of ^{133}Cs and K was not: asymmetry of ^{137}Cs frequency distributions is reported in other fungal species (Baeza et al., 2004; Gaso et al., 1998; Ismail, 1994). According to Gillett & Crout (2000), the frequency distribution of ^{137}Cs appears species dependent: high accumulating species tend to be normally distributed and low accumulating species tend to be log-normally distributed. However, lognormal distribution is almost the default for concentration of radionuclides and is unlikely to be a species-specific phenomenon, as it also occurs in soil concentrations, which implies normal distribution would not be expected, even if large set of samples were analyzed.

1.8 Mechanisms of ^{137}Cs and alkali metal uptake by fungi

Generally, little is known about the mechanisms involved in the uptake and retention of radionuclides by fungi. Studies of uptake mechanisms and affinity for alkali metals in fungi are scarce, but some results are reviewed by Rodríguez-Navarro (2000). Compared to plants, fungal fruit bodies can be characterized by high ^{137}Cs, ^{133}Cs and Rb concentrations and low calcium (Ca) and strontium (Sr) concentrations. In a laboratory experiment with the wood-inhabiting mushroom *Pleurotus ostreatus* (Fr.) Kummer Y-l (Terada et al., 1998), ^{137}Cs uptake by mycelia decreased with increasing of ^{133}Cs, K or Rb concentration in the media, and K uptake by mycelia decreased with increasing of ^{133}Cs concentration. In an experiment with pure cultures of mycorrhizal fungi (Olsen et al., 1990) some species had preference for Cs over K and in the experiments with yeast (Conway & Duggan, 1958), K had preference over Cs and the affinity for alkali metal uptake decreased in the order $K^+ < Rb^+ < Cs^+$ followed by Na^+ and Li^+, with a relative ratio of 100:42:7:4:0.5. Fungi (mycelium and sporocarps) have a higher affinity for uptake of Rb and K to Cs, and based on the CR values for fungal sporocarps (Table 3), alkali metal can be ranked in the order $Rb^+ > K^+ > Cs^+$, with a relative ratio of 100:57:32, which is within the range of 100:88:50 derived by Yoshida & Muramatsu (1998).

The affinity for an alkali metal depends on the nutritional status of the organism, which at least partly explains differences reported between field experiments and laboratory experiments with a good nutrient supply. The mycorrhizal species *Sarcodon imbricatus* was found to be the most efficient in accumulating K, Rb and Cs, which was in agreement with results obtained by Tyler (1982), where a mean CR for Rb in litter decomposing fungus *Collybia peronata* was reported to be 41, and the mean CR for Rb in *Amanita rubescens*, which is mycorrhizal with several tree species, was above 100. However, lower ^{40}K content for mycorrhizal species is reported by Römmelt et al. (1990), which means mycorrhizal species do not necessarily accumulate alkali metals more efficiently than saprotrophic ones.

Accumulation of stable and radioactive cesium by fungi is apparently species-dependent but is affected by local environmental conditions. According to de Meijer et al. (1988), the variation in concentrations of stable and radioactive cesium in fungi of the same species is generally larger than the variation between different species and the variation in ^{137}Cs levels within the same genet of *S. varegatus* is as large as within non-genet populations of the species (Dahlberg et al., 1997), suggesting both interspecific and intrapopulation variation in the uptake of K, Rb, stable ^{133}Cs and ^{137}Cs, and that their relationships can be explained by factors other than genotype identity (Vinichuk et al., 2011). There is about two orders of magnitude variation in Cs uptake, with the highest CR value in e.g. *S. imbricatus* (256) and the lowest in *Lactarius deterrimus* (2.6), although other studies (Seeger & Schweinshaut, 1981) report the highest accumulation of stable Cs is in *Cortinarius* sp.

2. Cs (^{137}Cs and ^{133}Cs), K and Rb in *Sphagnum* plants

2.1 Introduction

Peatlands are areas where remains of plant litter have accumulated under water-logging as a result of anoxic conditions and low decomposability of the plant material. They are generally nutrient-poor habitats, particularly temperate and boreal bogs in the northern hemisphere, in which peat formation builds a dome isolating the vegetation from the surrounding groundwater. Hence, bogs are ombrotrophic, i.e. all water and nutrient supply to the vegetation is from aerial dust and precipitation, resulting in an extremely nutrient-poor ecosystem often formed and dominated by peat mosses (*Sphagnum*). *Sphagnum*-dominated peatlands with some groundwater inflow (i.e. weakly minerotrophic 'poor fens') are almost as nutrient poor and acid as true bogs. *Sphagnum* plants absorb and retain substantial amounts of fallout-derived radiocesium, and some attention has been given to the transfer of the radioactive cesium isotope ^{137}Cs within raised bogs (Bunzl & Kracke, 1989; Rosén et al., 2009), and relatively high ^{137}Cs bioavailability to bog vegetation and mosses in particular are found (Bunzl & Kracke, 1989).

The transfer of ^{137}Cs within a peatland ecosystem is different from that in forest or on agricultural land. In soils with high clay content, there is low bioavailability and low vertical migration rate of radiocesium due to binding to some clay minerals (Cornell, 1993). In nutrient-poor but organic-matter-rich forest soils, the vertical migration rate of ^{137}Cs is also low, but bioavailability is often high, particularly for mycorrhizal fungi (Olsen et al., 1990; Vinichuk & Johansson, 2003; Vinichuk et al., 2004; 2005). In forests and pastures, extensive fungal mycelium counteracts the downward transport of ^{137}Cs by an upward translocation flux (Rafferty et al., 2000); this results in a slow net downward transport of ^{137}Cs in the soil profile.

In peatlands, ^{137}Cs appears to move through advection in peat water (review by Turetsky et al., 2004). Small amounts of clay mineral in the peat reduce Cs mobility (MacKenzie et al., 1997), but most *Sphagnum* peat is virtually clay mineral free organic matter. In wet parts of open peatlands that lack fungal mycelium, the downward migration of ^{137}Cs in the *Sphagnum* layers is expected to be faster than in forest soil and Cs is continuously translocated towards the growing apex of the *Sphagnum* shoots, where it is accumulated. Some attempts have been made to investigate whether ^{137}C is associated with essential biomacromolecules in mosses and to determine the ^{137}Cs distribution among intracellular moss compartments (Dragović et al., 2004).

The chemical behavior of radiocesium is expected to be similar to that of stable ^{133}Cs and other alkali metals, i.e. K, Rb, which have similar physicochemical properties. Moreover, stable ^{133}Cs usually provides a useful analogy for observing long-term variation and transfer parameters of ^{137}Cs in a specific environment, particularly in peatlands that are cut off from an input of stable Cs from the mineral soil. As the relationship between K and Rb in fungi is not clearly understood, whether Cs follows the same pathways as K in *Sphagnum* is also unclear.

Thus, the ^{137}Cs activity concentration and mass concentration of K, Rb and ^{133}Cs was analyzed within individual *Sphagnum* plants (down to 20 cm depth) growing on a peatland in eastern central Sweden and its distribution in the uppermost capitulum and subapical segments of *Sphagnum* mosses were compared to determine the possible mechanisms involved in radiocesium uptake and retention within *Sphagnum* plants.

Additionally, the isotopic (atom) ratios of ^{137}Cs/K, ^{137}Cs/Rb and ^{137}Cs/^{133}Cs within individual *Sphagnum* plants were recorded for determining the distribution of ^{137}Cs and alkali metal, and to obtain a better understanding of the uptake mechanisms and the biological behavior of ^{137}Cs in nutrient-poor *Sphagnum* dominated ecosystem. There are few studies on the influence of alkali metals (K, Rb, ^{133}Cs) on ^{137}Cs distribution and cycling processes in peatlands.

Plant species growing on peat have varying degree capacities for influencing uptake and binding of radionuclides, but no systematic study has covered all the dominant species of *Sphagnum* peatlands their competition for radionuclides and nutrients. The important role of *Sphagnum* mosses in mineral nutrient turnover in nutrient-poor ecosystems, in particular their role in ^{137}Cs uptake and binding, necessitates a clear understanding of the mechanisms involved.

The general aim was to gain better insight into mechanisms governing the uptake of both radionuclides (^{137}Cs) and stable isotopes of alkali metals (K, Rb, [133]Cs) by *Sphagnum* mosses. The specific aim was to compare the distribution of ^{137}Cs, K, Rb and [133]Cs in the uppermost capitulum and subapical segments of *Sphagnum* mosses to be able to discuss the possible mechanisms involved in radiocesium uptake and retention within *Sphagnum* plants. Most results obtained in this study are published in collaboration with Prof. H. Rydin (Vinichuk et al., 2010a).

2.2 Study area and methods
2.2.1 Study area

The study area was a small peatland (Palsjömossen) within a coniferous forest in eastern central Sweden, about 35 km NW of Uppsala (60°03′40″N, 17°07′47″E): the peatland area sampled was open and *Sphagnum*-dominated (Figure 4). A weak minerotrophic influence was indicated by the dominance of *Sphagnum papillosum*, and the presence of *Carex rostrata*, *Carex pauciflora* and *Menyanthes trifoliata* (fen indicators in the region). The area had scattered hummocks, mostly built by *Sphagnum fuscum*, and was dominated by dwarf-shrubs such as *Andromeda polifolia*, *Calluna vulgaris*, *Empetrum nigrum* and *Vaccinium oxycoccos*. Sampling was within a 25 m^{-2} low, flat 'lawn community' (Rydin & Jeglum, 2006) totally covered by *S. papillosum*, *S. angustifolium* and *S. magellanicum* with an abundant cover of *Eriophorum vaginatum*. The water table was generally less than 15 cm below the surface: surface water was pH 3.9–4.4 (June 2009).

2.2.2 Methods

Samples of individual *Sphagnum* shoots that held together down to 20 cm were randomly collected in 2007 (May and September) and 2008 (July, August and September). Thirteen samples of *Sphagnum* plants were collected and analyzed; three in 2007 and 10 sets in 2008. Each sample consisted of approximately 20–60 individual *Sphagnum* plants (mostly *S. papillosum*, in a few cases *S. angustifolium* or *S. magellanicum*). In the laboratory, the fresh, individual, erect and tightly interwoven *Sphagnum* plants were sectioned into 1 cm (0–10) or 2 cm (10–20 cm) long segments down to 20 cm from the growing apex. The ^{137}Cs activity concentrations were measured in fresh *Sphagnum* segments. Thereafter, the samples were dried at 40°C to constant weight and analyzed for K, Rb and [133]Cs.

The activity concentration (Bq kg^{-1}) of ^{137}Cs in plant samples was determined by calibrated HP Ge detectors. Statistical error due to the random process of decay ranged between 5 and 10%. Plant material was measured in different geometries filled up, except a few samples that contained about 1 g of dry material. All ^{137}Cs activity concentrations were recalculated to the sampling date and expressed on a dry mass basis. The analysis of *Sphagnum* segments for K, Rb and Cs was by a combination of ICP-AES and ICP-SFMS techniques at ALS Scandinavia AB. For K concentration determination, ICP-AES was used and for [133]Cs and Rb, ICP-SFMS was used. The detection limits were 200 mg kg^{-1} for K, 0.04 mg kg^{-1} for [133]Cs and 0.008 mg kg^{-1} for Rb. The isotopic (atom) ratio of ^{137}Cs/[133]Cs was calculated with Equations 1 and 2 (Chao et al., 2008). Relationships between K, Rb and [133]Cs concentrations in different *Sphagnum* segments were determined by Pearson correlation coefficients. All statistical analyses were with Minitab (© 2007 Minitab Inc.) software.

Fig. 4. The study area of peatland, Palsjömossen: *Sphagnum*-dominated bog.

2.3 Distribution of Cs (¹³⁷Cs and ¹³³Cs), K and Rb within *Sphagnum* plants

Concentration values of Cs (¹³⁷Cs and ¹³³Cs) and neighboring alkali counterparts K and Rb in different segments of plant provide information on differences in their uptake, distribution and relationships. The averaged ¹³⁷Cs activity concentrations in *Sphagnum* segments are presented in Figure 5a. Within the upper 10 cm from the capitulum, ¹³⁷Cs activity concentration in *Sphagnum* plants was about 3350 Bq kg⁻¹, with relatively small variations. Below 10-12 cm, the activity gradually declined with depth and in the lowest segments of *Sphagnum*, ¹³⁷Cs activity concentrations was about 1370 Bq kg⁻¹.

For individual samples, K concentrations ranged between 508 and 4970 mg kg⁻¹ (mean 3096); Rb ranged between 2.4 and 31.4 mg kg⁻¹ (mean 18.9) and ¹³³Cs ranged between 0.046 and 0.363 mg kg⁻¹ (mean 0.204): averaged concentrations of K, Rb and ¹³³Cs in *Sphagnum* segments are presented in Figure 5b. Concentrations of Rb and ¹³³Cs were constant in the upper 0-10 cm segments of *Sphagnum* moss and gradually declined in the lower parts of the plant length; whereas, the concentration of K decreased with increasing depth below 5 cm. Generally, the distribution of all three alkali metals was similar to ¹³⁷Cs, but with a weaker increase of Rb towards the surface. The ¹³⁷Cs activity concentrations had the highest coefficient of variation (standard deviation divided by the mean) in *Sphagnum* (43%). The coefficients of variation were 35% for K, 35% for Rb and 37% for ^{133}Cs concentrations.

Two important features should be mentioned when discussing distributions of K, Rb, ^{133}Cs and ^{137}Cs in a *Sphagnum*-dominated peatland. Firstly, this type of peatland is extremely nutrient-poor, where only a few plant and fungal species producing small fruit bodies can grow and no mycorrhiza, except ericoid mycorrhiza, exists. Secondly, the upper part of the stratigraphy is composed of living *Sphagnum* cells that selectively absorb mineral ions from the surrounding water, and the binding of K, Rb and ^{133}Cs can be at exchange sites both outside and inside the cell.

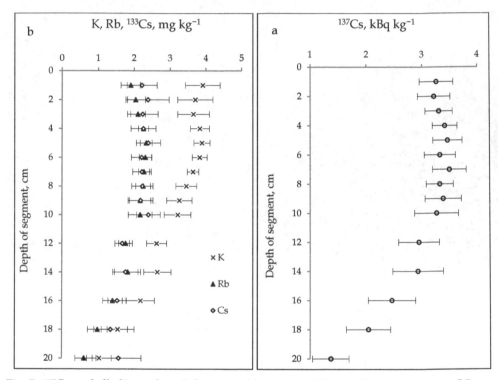

Fig. 5. ^{137}Cs and alkali metals in Sphagnum: (a) average ^{137}Cs activity concentration (kBq kg^{-1}) in *Sphagnum* segments (+/− SE, n = 13); (b) average concentrations of K (scale values should be multiplied by 10^3), Rb (x10^1) and ^{133}Cs (x10^{-1}) (mg kg^{-1}) in *Sphagnum* segments (+/− SE, n=4).

The distribution of ^{137}Cs within *Sphagnum* plants was similar to stable K, Rb and ^{133}Cs. The ^{137}Cs activity concentrations and K, Rb and ^{133}Cs concentrations were always highest in the uppermost 0-10 cm segments of *Sphagnum* (in the capitula and the subapical segments) and gradually decreased in older parts of plant. Such distribution could be interpreted as dependent on the living cells of capitula and living green segments in the upper part of *Sphagnum*. Similar patterns of K distribution within *Sphagnum* plants are reported (Hájek, 2008). ^{137}Cs is taken up and relocated by *Sphagnum* plants in similar ways to the stable alkali metals, as the ratios between K, Rb, Cs and ^{137}Cs in *Sphagnum* segments (Figure 6) were much the same down to about 16 cm, and displayed a slightly different pattern in the lower part of the plant.

2.4 Mass concentration and isotopic (atom) ratios between ^{133}Cs, K, Rb and ^{133}Cs, in segments of *Sphagnum* plants

Ratios between mass concentrations of all three alkali metals and ^{137}Cs activity concentrations, i.e. ^{133}Cs:^{137}Cs; K:^{137}Cs, Rb:^{137}Cs and ^{133}Cs:^{137}Cs, were constant through the upper part (0-16 cm) of *Sphagnum* plants (Figure 6). The ratio K/Rb was higher in the uppermost (0-2 cm) and the lowest (18-20 cm) parts of the plant (Figure 6).

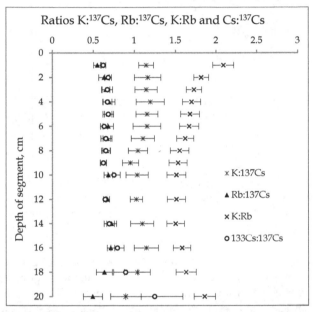

Fig. 6. Ratios between K:^{137}Cs, Rb:^{137}Cs (scale values should be multiplied by 10^{-2}), K:Rb (x10^2) and ^{133}Cs:^{137}Cs (x10^{-4}) in *Sphagnum* segments. Calculations based on concentrations in mg kg^{-1} for stable isotopes and Bq kg^{-1} for ^{137}Cs (+/− SE, n=13 for ^{137}Cs; n=4 for each of K, Rb and ^{133}Cs).

However, the isotopic (atom) ratios between ^{137}Cs activity concentrations and mass concentrations of alkali metals, i.e. ^{137}Cs/K, ^{137}Cs/Rb and ^{137}Cs/^{133}Cs, had distinctively different pattern of distribution through the upper part (0-20 cm) of *Sphagnum* plants (Figure 7). The ^{137}Cs/K ratio was relatively narrow through the upper part (0-16 cm) of *Sphagnum* plants and wider with increasing depth, whereas, the ^{137}Cs/^{133}Cs ratio was fairly constant through the upper part (0-12 cm) of *Sphagnum* plants and becomes narrower in the lower (14-20 cm) parts. The ^{137}Cs/Rb ratio was constant through the middle part (4-16 cm) of *Sphagnum* plants and somewhat narrower in the uppermost (0-4 cm) and lowest (16-20 cm) parts (Figure 7).

The distribution of the isotopic (atom) ratios between ^{137}Cs activity concentrations and mass concentrations of alkali metals K and Rb through the upper part (0-20 cm) of *Sphagnum* plants are probably conditioned by at least three processes: physical decay of ^{137}Cs atoms with time; attainment of equilibrium between stable [133]Cs and ^{137}Cs in the bioavailable fraction of peat soil; and, relation between cesium ([133]Cs and ^{137}Cs), K and Rb when taken up by the *Sphagnum* plant.

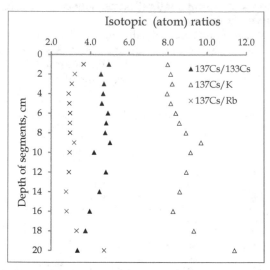

Fig. 7. Isotopic (atom) ratios ^{137}Cs/K (scale values should be multiplied by 10[-12]), ^{137}Cs/Rb (x 10[-09]), and ^{137}Cs/[133]Cs (x10[-07]) in *Sphagnum* segments. Calculations based on ^{137}Cs activity concentrations and mass concentrations of K, Rb [133]Cs (Eq. 2) (mean values, n=4 for each of ^{137}Cs, K, Rb and [133]Cs).

2.5 Relationships between [133]Cs, K, Rb and [133]Cs in segments of *Sphagnum* plants

Relationships between [133]Cs, K, Rb and [133]Cs in separate segments of *Sphagnum* plants is a tool allowing future investigate its uptake mechanism. There were close positive correlations between K, Rb and [133]Cs mass concentrations and ^{137}Cs activity concentrations in *Sphagnum* segments (Table 8). Correlation between ^{137}Cs activity concentrations and Rb mass concentrations (r=0.950; p=0.001) and correlation between K and Rb mass concentrations (r=0.952; p=0.001) in 10-20 cm length of *Sphagnum* plants were highest, but ^{137}Cs and K had a weaker correlation only when the upper 0-10 cm part of *Sphagnum* plants were analyzed (r=0.562; p=0.001). [137]Cs/[133]Cs isotope (atom) ratios and mass concentrations of alkali metals (K, Rb and [133]Cs) were not or negatively correlated (Table 8).

The marked decrease in ^{137}Cs activity concentration below 14 cm (Figure 5a) raises the question as to at what depth the 1986 Chernobyl horizon was when the sampling was done. A peat core was sampled in May 2003 at Åkerlänna Römosse, an open bog about 14 km SW of Pålsjömossen, by van der Linden et al. (2008). Detailed dating by ^{14}C wiggle-matching indicated the Chernobyl horizon was then at a depth of 17 cm. Depth-age data estimated a linear annual peat increment of 1.3 cm yr[-1] over the last decade (R[2]=0.998), indicating the Chernobyl horizon would be at about 23 cm deep when the ^{137}Cs sampling was done in 2007-08. Even if there are uncertainties in applying data from different peatlands, the Chernobyl horizon should be at, or below, the lowest segments sampled. Thus, an upward migration of ^{137}Cs was obvious, but no downward migration could be tested in the study.

The relatively unchanged ^{137}Cs/K, ^{137}Cs/Rb and ^{137}Cs/^{133}Cs isotopic (atom) ratios in the upper 0-14 cm part of Sphagnum plant and the noticeable widening below 14-16 cm supported this assumption. An upward migration of ^{137}Cs has been observed in earlier studies (Rosén et al., 2009); similarly, most ^{137}Cs from the nuclear bomb tests from 1963 was retained in the top few cm of *Sphagnum* peat 20 years after, but there was also a lower peak at the level where the 1963 peat was laid down (Clymo, 1983): *Cladonia* lichens also retain high activity concentrations in the shoot apices.

	^{137}Cs	K	Rb	^{133}Cs
0-10 cm length				
K	0.562***			
Rb	0.893***	0.632***		
^{133}Cs	0.840***	0.792***	0.802***	
^{137}Cs/^{133}Cs	–	−0.262	0.270	−0.157
10-20 cm length				
K	0.856***			
Rb	0.950***	0.952***		
^{133}Cs	0.645***	0.651***	0.664***	
^{137}Cs/^{133}Cs	–	0.122	0.219	−0.401

Table 8. Correlation coefficients between concentrations of potassium, rubidium and cesium (^{133}Cs and ^{137}Cs) in *Sphagnum* segments (*** $p=0.001$).

2.6 Mechanisms of ^{137}Cs and alkali metal uptake by *Sphagnum* plants

Presumably, ^{137}Cs is bound within capitula, living green segments and dead brown segments of *Sphagnum* plants. According to Gstoettner and Fisher (1997), the uptake of some metals (Cd, Cr, and Zn) in *Sphagnum papillosum* is a passive process as they living and dead moss accumulate metal equally. For a wide range of bryophytes, Dragović et al. (2004) found ^{137}Cs was primarily bound by cation exchange, with only a few percent occurring in biomolecules. *Sphagnum* mosses have remarkably high cation exchange capacity (Clymo, 1963), and according to Russell (1988), a high surface activity of *Sphagnum* is related to its high cation exchange capacity, which ranges between 90-140 meq/100 g. In a water saturated peat moss layer, water washes (1 L de-ionised water added to a column of about 1.4 L volume) removed about 60% of K from *Sphagnum* (Porter B. Orr, 1975), indicating this element was held on cation exchange sites. In turn, the desiccation of living moss usually causes cation leakage from cell cytoplasm, during which most of the effused K^+ is retained on the exchange sites and reutilized during recovery after rewetting (Brown & Brümelis, 1996; Bates, 1997).

However, this is not necessarily the case for ^{137}Cs, as ^{137}Cs has a weaker correlation with K, especially in the uppermost parts of the plant, which means ^{137}Cs uptake might be somewhat different from that of K. Even within the same segments of the plant, ^{137}Cs activity concentrations has higher variation than K concentration. An even stronger decoupling between ^{137}Cs and K is observed in the forest moss *Pleurozium schreberi*, in which ^{137}Cs is retained to a higher degree in senescent parts (Mattsson & Lidén, 1975). However, close correlations, were found between Rb and ^{137}Cs, which suggests similarities in their uptake and relocation: these observations complied with results reported for fungi (Vinichuk et al., 2010b; 2011).

Some lower parts of *Sphagnum* plants are still alive and able to create new shoots (Högström, 1997), however, although still connected to the capitulum, much of lower stem is dead. Thus, the decrease of ^{137}Cs activity concentration in plant segments below 10 cm indicates a release of the radionuclide from the dying lower part of *Sphagnum* and internal translocation to the capitulum.

The mechanism of radiocesium and alkali metal relocation within *Sphagnum* is probably the same active translocation as described for metabolites by Rydin & Clymo (1989). Although external buoyancy-driven transport (Rappoldt et al., 2003) could redistribute ^{137}Cs, field evidence suggests buoyancy creates a downward migration of K (Adema et al., 2006); thus, this mechanism appears unlikely. Likewise, a passive downwash and upwash (Clymo & Mackay, 1987) cannot explain accumulation towards the surface.

3. Conclusions from the Swedish studies

The concentrations of the three stable alkali elements K, Rb and ^{133}Cs and the activity concentration of ^{137}Cs were determined in various components of Swedish forests – bulk soil, rhizosphere, soil–root interface fraction, fungal mycelium and fungal sporocarps. The soil–root interface fraction was distinctly enriched with K and Rb, compared with bulk soil. Potassium concentration increased in the order bulk soil < rhizosphere < fungal mycelium < soil–root interface < fungal sporocarps, whereas, Rb concentration increased in the order bulk soil < rhizosphere < soil–root interface < fungal mycelium < fungal sporocarps.

Cesium was generally evenly distributed within bulk soil, rhizosphere and soil–root interface fractions, indicating no ^{133}Cs enrichment in these forest compartments.

The uptake of K, Rb and ^{133}Cs during the entire transfer process between soil and sporocarps occurred against a concentration gradient. For all three alkali metals, the levels of K, Rb and ^{133}Cs were at least one order of magnitude higher in sporocarps than in fungal mycelium.

Potassium uptake appeared to be regulated by fungal nutritional demands for this element and fungi had a higher preference for uptake of Rb and K than for Cs. According to their efficiency of uptake by fungi, the three elements may be ranked in the order $Rb^+ > K^+ > Cs^+$, with a relative ratio 100:57:32. Although the mechanism of Cs uptake by fungi could be similar to that of Rb, uptake mechanism for K appeared to be different. The variability in isotopic (atom) ratios of ^{137}Cs/K, ^{137}Cs/Rb and ^{137}Cs/^{133}Cs in the fungal sporocarps suggested they were independent on specific species of fungi. The relationships observed between concentration ratios ^{137}Cs/^{133}Cs and K, Rb and ^{133}Cs in fungal sporocarps also varied widely and were inconsistent. The concentration of K, Rb and ^{133}Cs in sporocarps appeared independent of the ^{137}Cs/^{133}Cs isotopic ratio.

The study of *S. variegatus* sporocarps sampled within 1 km^2 forest area with high ^{137}Cs fallout from the Chernobyl accident confirmed ^{133}Cs and ^{137}Cs uptake is not correlated with uptake of K; whereas, the uptake of Rb is closely related to the uptake of ^{133}Cs. Furthermore, the variability in ^{137}Cs and alkali metals (K, Rb and ^{133}Cs) among genotypes in local populations of *S. variegatus* is high and the variation appears to be in the same range as found in species collected at different localities. The variations in concentrations of K, Rb and ^{133}Cs and ^{137}Cs activity concentration in sporocarps of *S. variegatus* appear to be influenced more by local environmental factors than by genetic differences among fungal genotypes.

For *Sphagnum* the distribution of ^{137}Cs can be driven by several processes: cation exchange is important and gives similar patterns for monovalent cations; uptake/retention in living cells; and downwash and upwash by water outside the plants. However, the most important mechanism is internal translocation to active tissue and the apex, which can explain the accumulation in the top layer of the mosses.

4. Acknowledgements

The authors gratefully acknowledge the Swedish University of Agricultural Sciences (SLU), Sweden, for supporting the project. We would like to express our thanks to Dr. I. Nikolova for her assistance with the experiments and to the staff of the Analytica Laboratory, Luleå, Sweden, for ICP-AES *and* ICP-SFMS analyses. The project was financially supported by SKB (Swedish Nuclear Fuel and Waste Management Co).

5. References

Adema, E.; Baaijens, G.; van Belle, J.; Rappoldt, A.; Grootjans, A. & Smolders, A. (2006). Field evidence for buoyancy-driven water flow in a *Sphagnum* dominated peat bog. *Journal of Hydrology*, Vol.327, pp. 226– 234, ISSN 0022-1694

Baeza, A.; Hernández, S.; Guillén, F.; Moreno, J.; Manjón, J.L. & Pascual, R. (2004). Radiocaesium and natural gamma emitters in mushrooms collected in Spain. *Science of the Total Environment*, Vol.318, No.1-3, pp. 59-71, ISSN 0048-9697

Baeza, A.; Guillén, J.; Hernández, S.; Salas, A.; Bernedo, M.; Manjón, J. & Moreno, G. (2005). Influence of the nutritional mechanism of fungi (mycorrhize/saprophyte) on the uptake of radionuclides by mycelium. *Radiochimica Acta*, Vol.93, No.4, pp. 233-238, ISSN 0033-8230

Ban-nai T.; Yoshida S.; Muramatsu Y. & Suzuki A. (2005). Uptake of Radiocesium by Hypha of Basidiomycetes – Radiotracer Experiments. *Journal of Nuclear and Radiochemical Sciences*, Vol. 6, No.1, pp. 111-113, ISSN 1345-4749

Bates, J. (1997). Effects of intermittent desiccation on nutrient economy and growth of two ecologically contrasted mosses. *Annals of Botany*, Vol.79, pp.299–309, ISSN 0305-7364

Brown, D. & Brümelis, G. (1996). A biomonitoring method using the cellular distribution of metals in moss. *Science of the Total Environment*, Vol.187, pp. 153–161, ISSN 0048-9697

Brown, G. & Cummings, S. (2001). Potassium uptake and retention by *Oceanomonas baumannii* at low water activity in the presence of phenol. *FEMS Microbiology Letters*, Vol.205, No.1, pp. 37–41, ISSN 0378-1097

Brunner, I.; Frey, B. & Riesen, T. (1996). Influence of ectomycorrhization and cesium/potassium ratio on uptake and localization of cesium in Norway spruce seedlings. *Tree Physiology*, Vol.16, pp. 705-711, ISSN 0829-318X

Bunzl, K. & Kracke, W. (1989). Seasonal variation of soil-to-plant transfer of K and fallout ^{134,137}Cs in peatland vegetation. *Health Physics*, Vol.57, pp. 593-600, ISSN 0017-9078

Bystrzejewska-Piotrowska, G. & Bazala, M. (2008). A study of mechanisms responsible for incorporation of cesium and radiocaesium into fruitbodies of king oyster mushroom (*Pleurotus eryngii*). *Journal of Environmental Radioactivity*, Vol.99, pp. 1185-1191, ISSN 0265-931X

Chao J.; Chiu, C. & Lee, H. (2008). Distribution and uptake of ^{137}Cs in relation to alkali metals in a perhumid montane forest ecosystem. *Applied Radiation and Isotopes*, Vol.66, pp. 1287-1294, ISSN 0969-8043

Cornell, R. (1993). Adsorption of cesium on minerals: A review. *Journal of Radioanalytical and Nuclear Chemistry*, Vol.171, pp. 483-500, ISSN 0236-5731

Clymo, R. (1963). Ion exchange in Sphagnum and its relation to bog ecology. *Annals of Botany*, (Lond.) Vol.27, pp. 309-324, ISSN 0305-7364

Clymo, R. (1983). Peat. In: Gore, A.J.P. (Ed.), Ecosystems of the world. 4A. Mires: swamp, bog, fen and moor. General studies, 159-224, ISBN 0-444-42003-7, Elsevier, Amsterdam.

Clymo, R.; & Mackay, D. (1987). Upwash and downwash of pollen and spores in the unsaturated surface layer of *Sphagnum*-dominated peat. *New Phytologist*, Vol.105, pp. 175-183, ISSN 0028-646X

Conway, E.; & Duggan, F. (1958). A cation carrier in the yeast cell wall. *Biochemistry Journal*, Vol.69, pp. 265-274, ISSN 0006-2960

Dahlberg, A.; Nikolova, I.; Johanson, K. (1997). Intraspecific variation in ^{137}Cs activity concentration in sporocarps of *Suillus variegatus* in seven Swedish populations. *Mycological Research*, Vol.101, pp.545-551, ISSN 0953-7562

de Meijer, R.; Aldenkamp, F. & Jansen, A. (1988). Resorption of cesium radionuclides by various fungi. *Oecologia*, Vol.77, pp. 268-272, ISSN 0029-8549

Dighton, J.; Clint, G. & Poskitt J. (1991). Uptake and accumulation of ^{137}Cs by upland grassland soil fungi: a potential pool of Cs immobilisation. *Mycological Research*, Vol.95, No.9, pp. 1052–1056, ISSN 0953-7562

Dragović, S.; Nedić, O.; Stanković, S. & Bačić, G. (2004). Radiocaesium accumulation in mosses from highlands of Serbia and Montenegro: chemical and physiological aspects. *Journal of Environmental Radioactivity*, Vol.77, pp. 381-388, ISSN 0265-931X

Drobner, U. & Tyler G. (1998). Conditions controlling relative uptake of potassium and rubidium by plants from soils. *Plant and Soil*, Vol.201, pp. 285–293, ISSN 0032-079X

Enghag, P. (2000). *Jordens grundämnen och deras upptäckt*. ISBN 9789175485904, Industrrilitteratur. Stockholm.

Gaso, M.; Segovia, N.; Morton, O.; Cervantes, M.; Godinez, L.; Peña, P. & Acosta, E. (2000). [137]Cs and relationships with major and trace elements in edible mushrooms from Mexico. *Science of the Total Environment*, Vol.262, No.1-2, pp. 73–89, ISSN 0048-9697

Gillett, A. & Crout, N. (2000). A review of ^{137}Cs transfer to fungi and consequences for modeling environmental transfer. *Journal of Environmental Radioactivity*, Vol.48, pp. 95-121, ISSN 0265-931X

Gorban, G. & Clegg, S. (1996). A conceptual model for nutrient availability in the mineral soil-root system. *Canadian Journal of Soil Science*, Vol.76, pp. 125–131, ISSN 1918-1841

Gstoettner, E. & Fisher, N. (1997). Accumulation of cadmium, chromium, and zinc by the moss *Sphagnum papillosum* Lindle. *Water, Air, Soil Pollution*, Vol.93, pp. 321-330, ISSN 0049-6979

Guillitte, O.; Melin, J. & Wallberg, L. (1994). Biological pathways of radionuclides originating from the Chernobyl fallout in a boreal forest ecosystem. *Science of the Total Environment*, Vol.157, pp. 207–215, ISSN 0048-9697

Gyuricza, V.; Dupré de Boulois, H. & Declerck, S. (2010). Effect of potassium and phosphorus on the transport of radiocaesium by arbuscular mycorrhizal fungi. *Journal of Environmental Radioactivity*, Vol.101, pp. 482–487, ISSN 0265-931X

Hájek, T. (2008). Ecophysiological adaptations of coexisting *Sphagnum* mosses. PhD. thesis. University of South Bohemia, Faculty of Science, Czech Republic, 98 pp.

Högström, S. (1997). Habitats and increase of Sphagnum in the Baltic Sea island Gotland, Sweden. *Lindbergia*, Vol.22, pp. 69-74, ISSN 0105-0761

Horyna, J. & Řanad, Z. (1988). Uptake of radiocaesium and alkali metals by mushrooms. *Journal of Radioanalytical and Nuclear Chemistry*, Vol.127, pp. 107–120, ISSN 0236-5731

Ismail, S. (1994). Distribution of Na, K, Rb, Cs and ^{137}Cs in some Austrian higher fungi. *Biology of Trace Element Research*, Vol.43-45, No.1, pp. 707–714, ISSN 01634984

Johanson, K.J. &, Bergström, R. (1994). Radiocaesium transfer to man from moose and roe deer in Sweden. *Science of the Total Environment*, Vol.157, pp. 309-316, ISSN 0048-9697

Kalač P. (2001). A review of edible mushroom radioactivity. *Food Chemistry*, Vol.75, pp. 29–35, ISSN 0308-8146

Kammerer, L.; Hiersche, L. & Wirth, E. (1994). Uptake of radiocaesium by different species of mushrooms. *Journal of Environmental Radioactivity*, Vol.23, pp. 135-150, ISSN 0265-931X

Karadeniz, Ö. & Yaprak, G. (2007). Dynamic equilibrium of radiocesium with stable cesium within the soil–mushroom system in Turkish pine forest. *Environmental Pollution*, Vol.148, No.1, pp. 316-324, ISSN 0269-7491

Kuwahara, C.; Watanuki, T.; Matsushita, K.; Nishina, M. & Sugiyama, H. (1998). Studies on uptake of cesium by mycelium of the mushroom *(Pleurotus ostreatus)* by ^{133}Cs-NMR. *Journal of Radioanalytical and Nuclear Chemistry*, Vol.235, No.1-2, pp. 191-194, ISSN 0236-5731

Leake, J. & Read, D. (1997). Mycorrhizal fungi in terrestrial habitats. Chapter 18. In: *The Mycota, Vol.IV: Environmental and Microbial Relationship*, D.T. Wicklow & B. Söderström, (Ed.), 281-301, ISBN 978-3540-71839-0, Springer Verlag, Berlin

MacKenzie, A.; Farmer, J. & Sugden, C. (1997). Isotopic evidence of the relative retention and mobility of lead and radiocaesium in Scottish ombrotrophic peats. *Science of the Total Environment*, Vol.203, pp. 115-127, ISSN 0048-9697

Marschner, H. (1995). *Mineral nutrition of higher plants*. Academic press, 2nd ed, ISBN 0-12-473543-6, London, UK

Mascanzoni, D. (2009). Long-term transfer of ^{137}Cs from soil to mushrooms in a semi-natural environment. *Journal of Radioanalytical and Nuclear Chemistry*, Vol.282, pp. 427-431, ISSN 0236-5731

Mattsson, S. & Lidén, K. (1975). ^{137}Cs in carpets of the forest moss Pleurozium schreberi, 1961-1973. *Oikos*, Vol.26, pp. 323-327, ISSN 0030-1299

McGee, E.; Synnott, H.; Johanson, K.; Fawaris, B.; Nielsen, S. Horrill, A.D., et al., (2000). Chernobyl fallout in a Swedish spruce forest ecosystem. *Journal of Environmental Radioactivity*, Vol.48, pp. 59-78, ISSN 0265-931X.

Myttenaere, C.; Schell, W.; Thiry, Y.; Sombre, L.; Ronneau, C.; van der Stegen de Schriek, J. (1993). Modeling of Cs-137 cycling in forests: recent developments and research needed. *Science of the Total Environment*, Vol.136, pp. 77–91, ISSN 0048-9697

Nikolova, I.; Johanson, K. & Dahlberg, A. (1997). Radiocaesium in fruitbodies and mycorrhizae in ectomycorrhizal fungi. *Journal of Environmental Radioactivity*, Vol.37, pp. 115–125, ISSN 0265-931X

Nyholm, N. & Tyler G. (2000). Rubidium content of plant, fungi and animals closely reflects potassium and acidity conditions of forest soils. *Forest Ecology and Management*, Vol.134, pp. 89–96, ISSN 0378-1127

Olsen, R.; Joner, E. & Bakken. L. (1990). Soil fungi and the fate of radiocaesium in the soil ecosystem - a discussion of possible mechanisms involved in the radiocaesium accumulation in fungi, and the role of fungi as a Cs-sink in the soil. In: G. Desmet, P. Nassimbeni, & M. Belli (Eds.), *Transfer of radionuclides in natural and semi-natural environment*, pp. 657–663, ISBN 9781851665396, Luxemburg: Elsevier Applied Science

Parekh, N.; Poskitt, J.; Dodd, B.; Potter, E. & Sanchez, A. (2008). Soil microorganisms determine the sorption of radionuclides within organic soil systems. *Journal of Environmental Radioactivity*, Vol.99, pp. 841-852, ISSN 0265-931X

Porter B. Orr. 1975. Available from http://scholar.lib.vt.edu/ejournals/JARS/v30n3/v30n3-orr.htm

Rafferty, B.; Dawson, D, & Kliashtorin, A. (1997). Decomposition in two pine forests: the mobilisation of ^{137}Cs and K from forest litter. *Soil Biology and Biochemistry*, Vol.29, No.11/12, pp. 1673–1681, ISSN 0038-0717

Rafferty, B.; Brennan, M.; Dawson, D. & Dowding, D. (2000). Mechanisms of ^{137}Cs migration in coniferous forest soils. *Journal of Environmental Radioactivity*, Vol.48, pp. 131-143, ISSN 0265-931X

Rappoldt, C.; Pieters, G.; Adema, E.; Baaijens, G.; Grootjans, A. & van Duijn, C. (2003). Buoyancy-driven flow in peat bogs as a mechanism for nutrient recycling. *Proceedings of the National Academy of Sciences of the United States of America*, Vol.100, pp. 14937-14942, ISSN 1091-6490

Read, D.; Perez-Moreno, J. (2003). Mycorrhizas and nutrient cycling in ecosystems – a journey towards relevance? *New Phytologist*, Vol.157, pp. 475–492, ISSN 0028-646X

Rodríguez-Navarro, A. (2000). Potassium transport in fungi and plants. *BBA - Biochimica et Biophysica Acta*, Vol.1469, pp. 1–30, ISSN 1388-1981

Römmelt, R.; Hiersche, L.; Schaller, G. & Wirth, E. (1990). Influence of soil fungi (basidiomycetes) on the migration of ^{134}Cs+^{137}Cs and ^{90}Sr in coniferous forest soils. In: G. Desmet, P. Nassimbeni and M. Belli, Editors, *Transfer of radionuclides in natural and seminatural environments*, pp. 152–160, ISBN 9781851665396, Elsevier, London

Rosén, K.; Vinichuk, M. & Johanson, K. (2009). ^{137}Cs in a raised bog in central Sweden. Journal of Environmental Radioactivity, Vol.100, pp. 534–539, ISSN 0265-931X

Rosén, K.; Vinichuk, M.; Nikolova, I. & Johanson, K. (2011). Long-term effects of single potassium fertilization on ^{137}Cs levels in plants and fungi in a boreal forest ecosystem. *Journal of Environmental Radioactivity*, Vol.102, pp. 178-184, ISSN 0265-931X

Rühm, W.; Kammerer, L.; Hiersche, L. & Wirth, E. (1997). The ^{137}Cs/^{134}Cs ratio in fungi as an indicator of the major mycelium location in forest soil. *Journal of Environmental Radioactivity*, Vol.35, pp. 129–148, ISSN 0265-931X

Russell, E. (1988). *Soil conditions and plants growth*, Eleventh Edition, Longmans, ISBN 0582446775, London, England

Rydin, H. & Clymo, R. (1989). Transport of carbon and phosphorus about Sphagnum. *Proceedings of Royal Society*, Biol. Sci., Vol.237, pp. 63-84, ISSN 1471-2954, London

Rydin, H. & Jeglum, J. (2006). *The biology of peatlands*, ISBN 978-0-19-852871-5, Oxford University Press

Seeger, R. & Schweinshaut, P. (1981). Vorkommen von caesium in höheren pilzen. *Science of the Total Environment*, Vol.19, pp. 253–276, ISSN 0048-9697

Skuterud, L.; Travnicova, I.; Balonov, M.; Strand, P. & Howard, B. (1997). Contribution of fungi to radiocaesium intake by rural populations in Russia. *Science of the Total Environment*, Vol.193, No.3, pp. 237–242, ISSN 0048-9697

Smith, S. & Read, D. (1997). *Mycorrhizal symbiosis* (2nd ed.), ISBN 0-12-652840-3, London: Academic Press

Steiner, M.; Linkov, I. & Yoshida, S. (2002). The role of fungi in the transfer and cycling of radionuclides in forest ecosystems. *Journal of Environmental Radioactivity*, Vol.58, pp. 217–241, ISSN 0265-931X

Tanesaka, E.; Masuda, H. & Kinugawa K. (1993). Wood degrading ability of basidiomycetes that are wood decomposers, litter decomposers, or mycorrhizal symbionts. *Mycologia*, Vol.85, pp. 347–354, ISSN 0027-5514

Terada, H.; Shibata, H.; Kato, F. & Sugiyama, H. (1998). Influence of alkali elements on the accumulation of radiocesium by mushrooms. *Journal of Radioanalytical and Nuclear Chemistry*, Vol.235, No.1-2, pp. 195–200, ISSN 0236-5731

Tsukada, H.; Takeda, A.; Hisamatsu, S. & Inaba, J. Inequilibrium between Fallout ^{137}Cs and Stable Cs in Cultivated Soils. Radionuclides in Soils and Sediments, and their Transfer to Biota, In: *Proceedings of 18th World Congress of Soil Science*, 15 July 2006 Available   from http://crops.confex.com/crops/wc2006/techprogram/P17531.HTM

Turetsky, M.; Manning, S. & Wieder, R. (2004). Dating recent peat deposits. *Wetlands*, Vol.24, pp. 324-356, ISSN 0277-5212

Tyler, G. (1982). Accumulation and exclusion of metals in *Collybia peronata* and *Amaniuta rubescens*. *Transactions of British Mycological Society*, Vol.79, pp. 239–241, ISSN 0007-1536

van der Linden, M.; Vickery, E.; Charman, D. & van Geel, B. (2008). Effects of human impact and climate change during the last 350 years recorded in a Swedish raised bog deposit. *Palaeogeography, Palaeoclimatology, Palaeoecology*, Vol.262, pp. 1-31, ISSN 0031-0182

Vinichuk, M. & Johanson, K. (2003). Accumulation of ^{137}Cs by fungal mycelium in forest ecosystems of Ukraine. *Journal of Environmental Radioactivity*, Vol.64, pp. 27-43, ISSN 0265-931X

Vinichuk, M.; Johanson, K. & Taylor, A. (2004). ^{137}Cs in the fungal compartment of Swedish forest soils. *Science of the Total Environment*, Vol.323, pp. 243–251, ISSN 0048-9697

Vinichuk, M.; Johanson, K.; Rosén, K. & Nilsson, I. (2005). Role of fungal mycelium in the retention of radiocaesium in forest soils. *Journal of Environmental Radioactivity*, Vol.78, pp. 77–92, ISSN 0265-931X

Vinichuk, M.; Johanson, K.; Rydin, H. & Rosén, K. (2010a). The distribution of ^{137}Cs, K, Rb and Cs in plants in a Sphagnum-dominated peatland in eastern central Sweden. *Journal of Environmental Radioactivity*, Vol.101, pp. 170–176, ISSN 0265-931X

Vinichuk, M.; Taylor, A.; Rosén, K. & Johanson, K. (2010b). Accumulation of potassium, rubidium and cesium ([133]Cs and ^{137}Cs) in various fractions of soil and fungi in a Swedish forest. *Science of the Total Environment*, Vol.408, pp. 2543–2548, ISSN 0048-9697

Vinichuk, M.; Rosén, K.; Johanson, K. & Dahlberg, A. (2011). Correlations between potassium, rubidium and cesium ([133]Cs and ^{137}Cs) in sporocarps of *Suillus variegatus* in a Swedish boreal forest. *Journal of Environmental Radioactivity*, Vol.102, No.4, pp. 386-392, ISSN 0265-931X

Wallace, A. (1970). Monovalent–ion carrier effects on transport of ^{86}Rb and ^{137}Cs into bush bean plants. *Plant and Soil*, Vol.32, pp. 526–520, ISSN 0032-079X

White, P. & Broadley, M. (2000). Mechanisms of caesium uptake by plants. *New Phytologist*, Vol.147, pp. 241-256, ISSN 0028-646X

Yoshida, S. & Muramatsu, Y. (1998). Concentration of alkali and alkaline earth elements in mushrooms and plants collected in a Japanese pine forest, and their relationship with [137]Cs. *Journal of Environmental Radioactivity*, Vol.41, No.2, pp. 183–205, ISSN 0265-931X

Permissions

The contributors of this book come from diverse backgrounds, making this book a truly international effort. This book will bring forth new frontiers with its revolutionizing research information and detailed analysis of the nascent developments around the world.

We would like to thank Nirmal Singh, for lending his expertise to make the book truly unique. He has played a crucial role in the development of this book. Without his invaluable contribution this book wouldn't have been possible. He has made vital efforts to compile up to date information on the varied aspects of this subject to make this book a valuable addition to the collection of many professionals and students.

This book was conceptualized with the vision of imparting up-to-date information and advanced data in this field. To ensure the same, a matchless editorial board was set up. Every individual on the board went through rigorous rounds of assessment to prove their worth. After which they invested a large part of their time researching and compiling the most relevant data for our readers. Conferences and sessions were held from time to time between the editorial board and the contributing authors to present the data in the most comprehensible form. The editorial team has worked tirelessly to provide valuable and valid information to help people across the globe.

Every chapter published in this book has been scrutinized by our experts. Their significance has been extensively debated. The topics covered herein carry significant findings which will fuel the growth of the discipline. They may even be implemented as practical applications or may be referred to as a beginning point for another development. Chapters in this book were first published by InTech; hereby published with permission under the Creative Commons Attribution License or equivalent.

The editorial board has been involved in producing this book since its inception. They have spent rigorous hours researching and exploring the diverse topics which have resulted in the successful publishing of this book. They have passed on their knowledge of decades through this book. To expedite this challenging task, the publisher supported the team at every step. A small team of assistant editors was also appointed to further simplify the editing procedure and attain best results for the readers.

Our editorial team has been hand-picked from every corner of the world. Their multi-ethnicity adds dynamic inputs to the discussions which result in innovative outcomes. These outcomes are then further discussed with the researchers and contributors who give their valuable feedback and opinion regarding the same. The feedback is then collaborated with the researches and they are edited in a comprehensive manner to aid the understanding of the subject.

Apart from the editorial board, the designing team has also invested a significant amount of their time in understanding the subject and creating the most relevant covers. They scrutinized every image to scout for the most suitable representation of the subject and create an appropriate cover for the book.

The publishing team has been involved in this book since its early stages. They were actively engaged in every process, be it collecting the data, connecting with the contributors or procuring relevant information. The team has been an ardent support to the editorial, designing and production team. Their endless efforts to recruit the best for this project, has resulted in the accomplishment of this book. They are a veteran in the field of academics and their pool of knowledge is as vast as their experience in printing. Their expertise and guidance has proved useful at every step. Their uncompromising quality standards have made this book an exceptional effort. Their encouragement from time to time has been an inspiration for everyone.

The publisher and the editorial board hope that this book will prove to be a valuable piece of knowledge for researchers, students, practitioners and scholars across the globe.

List of Contributors

Raad Obid Hussain and Hayder Hamza Hussain
College of Science/Kufa University, Iraq

Sun-Chan Jeong
Institute of Particle and Nuclear Studies (IPNS), High Energy Accelerator Research Organization (KEK) 1-1 Oho, Japan

A. M. Saliba-Silva, E. F. Urano de Carvalho and M. Durazzo
Nuclear Fuel Center of Nuclear and Energy Research Institute, Brazilian Commission of Nuclear Energy, São Paulo, Brazil

H. G. Riella
Chemical Engineering Department of University of Santa Catarina, Florianopólis, Brazil

İbrahim Han
Ağrı İbrahim Çeçen University, Turkey

Sevil Porikli
Erzincan University, Faculty of Art and Sciences, Department of Physics, Turkey

Yakup Kurucu
Atatürk University, Faculty of Sciences, Department of Physics, Turkey

Shinsuke Mori
NARO Western Region Agricultural Research Center, Japan

Akira Kawasaki, Satoru Ishikawa and Tomohito Arao
National Institute for Agro-Environmental Sciences, Japan

B. de Celis, V. del Canto, R. de la Fuente and J.M. Lumbreras
Escuela de Ingenieria Industrial, Universidad de Leon, Spain

J. Mundo
Universidad Autonoma de Puebla, Mexico

B. de Celis Alonso
Universidad de Erlangen, Germany

Thanasis E. Economou
Laboratory for Astrophysics and Space Research, Enrico Fermi Institute, University of Chicago, Chicago, USA

E. Tel, M. Sahan, H. Sahan and F. A.Ugur
Faculty of Arts and Science, Osmaniye Korkut Ata University, Turkey

A. Aydin
Faculty of Arts and Science, Kirikkale University, Turkey

A. Kaplan
Faculty of Arts and Science, Süleyman Demirel University, Turkey

Klas Rosén
Department of Soil and Environment, Swedish University of Agricultural Sciences, Sweden

Anders Dahlberg
Department of Forest Mycology and Pathology, Swedish University of Agricultural Sciences, Sweden

Mykhailo Vinichuk
Department of Soil and Environment, Swedish University of Agricultural Sciences, Sweden
Department of Ecology, Zhytomyr State Technological University, Ukraine

Printed in the USA
CPSIA information can be obtained
at www.ICGtesting.com
JSHW011408221024
72173JS00003B/459

9 781632 380524